Helen Wheatley is Associate Professor in Film and Television Studies at the University of Warwick, UK. She is editor of *Re-viewing Television History: Critical Issues in Television Historiography* (I.B.Tauris, 2007), co-editor of *Television for Women: New Directions* (2016) and author of *Gothic Television* (2006). Her research focuses on television history and aesthetics.

'This superb book is no small accomplishment. *Spectacular Television* provides a painstaking archaeology of televisual excess as both a formal and industrial practice and places television in dialogue with an array of illuminating approaches. In doing so, Wheatley develops a precedent-setting research model and absolute must-read for television studies students and scholars everywhere.'
— Professor John T. Caldwell, UCLA, USA

'Helen Wheatley's innovative *Spectacular Television* exposes our contemporary oversight of the significance of television's visual strategies. Through a series of compelling case studies, and a fascinating history of television as technology, she demonstrates how television — the apparent 'poor relation' of cinema in terms of spectacle — always enticed and continues to entrance audiences with images that variously provoke curiosity, wonder and disgust.'
— Professor Karen Lury, University of Glasgow

'At last a book that really understands the "vision" part of television, and television studies. Helen Wheatley here rediscovers responses to the wide range of visual delights — and shocks! — that the medium has offered since its inception, highlighting television' "spectacularity" through careful analyses.'
— Professor John Ellis, Royal Holloway, University of London

HELEN WHEATLEY

SPECTACULAR TELEVISION
EXPLORING TELEVISUAL PLEASURE

Published in 2016 by
I.B.Tauris & Co. Ltd
London • New York
www.ibtauris.com

Copyright © Helen Wheatley

The right of Helen Wheatley to be identified as the author of this work has been asserted by the author in accordance with the Copyright, Designs and Patents Act 1988.

All rights reserved. Except for brief quotations in a review, this book, or any part thereof, may not be reproduced, stored in or introduced into a retrieval system, or transmitted, in any form or by any means, electronic, mechanical, photocopying, recording or otherwise, without the prior written permission of the publisher.

Every attempt has been made to gain permission for the use of the images in this book. Any omissions will be rectified in future editions.

References to websites were correct at the time of writing.

International Library of the Moving Image 23

HB ISBN: 978 1 78076 736 9
PB ISBN: 978 1 78076 737 6
eISBN: 978 1 78672 096 2
ePDF: 978 1 78673 096 1

A full CIP record for this book is available from the British Library
A full CIP record is available from the Library of Congress

Library of Congress Catalog Card Number: available

Printed and bound in Great Britain by T.J. International, Padstow, Cornwall

Contents

List of illustrations   vii
Acknowledgements   ix
Author's note   xii

Introduction: What is spectacular television? What is (tele)visual pleasure?   1

PART I: SPECTACULAR HISTORIES, SPECTACULAR TECHNOLOGIES

1   Television comes to town: The spectacle of television at the mid-twentieth-century exhibition and beyond   23

2   Spectacular colour? Reconsidering the launch of colour television in Britain   56

PART II: SPECTACULAR LANDSCAPES AND THE NATURAL WORLD: EXPLORING BEAUTIFUL TELEVISION

3   At home on safari: Colonial spectacle, domestic space and 1950s television   79

4   Visual pleasure, natural history television and televisual beauty   98

5   Television's landscapes, (tele)visual pleasure and the imagined elsewhere   122

PART III: SPECTACULAR BODIES AND (TELE)VISUAL PLEASURE

6   Fascinating bodies: Looking inside television's somatic spectacle   153

7   The erotics of television   190

Conclusion: Sites of wonder, sights of wonder	223
Notes	**232**
Bibliography	**244**
Index	**265**

List of Illustrations

1.1. Cossor Electronics advertisement from the Festival of Britain catalogue (Ian Cox (1951) *The South Bank Exhibition: A Guide to the Story It Tells*, London: HMSO: xix). 27
1.2. Ekco advert from the Festival of Britain catalogue (Ian Cox (1951) *The South Bank Exhibition: A Guide to the Story It Tells*, London: HMSO: xxviii). 28
1.3. The Telekinema and the Television Pavilion at the Festival of Britain (1951) (Architectural Press Archive/RIBA Collections). 31
1.4. Inside the Telekinema at the Festival of Britain (1951) (John Maltby/RIBA Collections). 33
1.5. The television personality as lifestyle branding at the Ideal Home Show, 2015: Laurence Llewellyn Bowen. 53
1.6. The television personality as lifestyle branding at the Ideal Home Show, 2015: Gregg Wallace. 54
1.7. The Virgin Media 'Telly Heaven' Smart Home at the Ideal Home Show, 2015. 55
3.1. The colonial view: *Filming Wild Animals* (BBC, 1954). 91
3.2. The colonial view: *On Safari* (BBC, 1957). 96
4.1. The artistically mutilated body: *Hannibal* (NBC, 2013–15). 117
4.2. An elegy for beautiful television? *Frozen Planet* (BBC1, 2011). 121
5.1. Television reconfigured as wall-hanging art (Coventry, UK, 2009). 125
5.2. The framed view of intentional landscape: *Britain's Favourite View* (ITV, 2007). 130
5.3. A hesitation on intentional landscape: *Wainwright Walks* (Skyworks for BBC4, 2007). 132
6.1. Body horror in *Incredible Human Machine* (National Geographic, 2007). 168

6.2. Gunther von Hagens with cadaver in *Anatomy for Beginners* (Channel 4, 2005). 184
7.1. The ordinary spectacle of sex in *Game of Thrones* (HBO, 2011–). 198
7.2. The object of the brazen gaze in *Poldark* (BBC1, 2015–). 216
7.3. Symmetrical composition and the desirable/desiring body in *Outlander* (Starz, 2014–). 219
C.1. Big-screen television viewing at the University of Warwick (photo courtesy of Sarah Wall). 227

Acknowledgements

This book has been a very long time in the writing and as such, a great number of people have played a part in its genesis, development and completion. All of my colleagues in the Department of Film and Television Studies at the University of Warwick, including the successive heads of department who helped me to find time to write various sections of the book, have aided me along the way; snatched conversations in the corridor and exchanges of emails and ideas have made this a much richer work. I am particularly grateful to Charlotte Brunsdon for her encouragement and for helping me to rethink my structure at a critical moment, and to all the other members of the Midlands Television Research Group for listening to my ideas about television spectacle over the last ten years. I have greatly valued talking about this work with Sofia Bull, as well as with the undergraduate students in the third-year module 'Television and Audio Visual Cultures', and my brilliant TV-focused PhD students Tom Steward, Charlotte Stevens and Zöe Shacklock, who have all taught me a great deal about television.

My friend and colleague Rachel Moseley has read every part of this book as it has slowly developed; she has copy-edited, pushed and challenged my ideas, and encouraged and supported me, and it is no exaggeration to say I couldn't have done it without her. I am most fortunate to have worked with Rachel and with Helen Wood, both brilliant television scholars and brilliant friends, on other research projects during the writing of this work. They have both enriched the ideas in this work. I am also immensely grateful to my friends and colleagues Tracey McVey, Anne Birchall and Heather Hares for their support and good humour throughout the research period.

A book which is partly about television history cannot be written without the support of a whole host of archivists and librarians: I have been particularly grateful for the help of Jess Hogg at the BBC Written Archives,

Andy O'Dwyer and Tony Ageh at the BBC, the staff of the Archive of Art and Design at Blythe House, and Richard Perkins in the University of Warwick library (who is a master at spotting relevant reading in obscure journals and is generally the fount of all knowledge). I am also indebted to everyone who gave up time to participate in the audience research which lies at the heart of several chapters in this book, from the people who discussed their picnics with me at *The Blue Planet Prom in the Park* to those who were absolutely candid about their erotic relationship with television. I didn't blush much.

As with any research project, being invited to present your ideas at critical stages in their development has a huge impact on the work. I am indebted to the following colleagues and all attendees who listened and responded to my work at the events where invited me to speak: Mandy Merck and James Bennett at Royal Holloway; James Fenwick, Laura Mee and Johnny Walker at De Montfort University; Yelda Özkoçak at Uşak University; Alex Clayton and Sarah Street at the University of Bristol; Lisa Taylor at Leeds Metropolitan University; Therese Davis and Adrian Martin at Monash University; Anna Reynolds Cooper at the University of Warwick; Joanne Hollows at Nottingham Trent University; Nicky North at the BFI; Lynn Spigel and Max Dawson at Northwestern University; and Jamie Sexton at the University of Wales, Aberystwyth. Some of the chapters in this book are based on previously republished work: my thanks therefore go to Ann Gray and the editors/reviewers at the *European Journal of Cultural Studies,* Duncan Petrie and the editors/reviewers at the *Journal of British Cinema and Television,* Karen Lury and editors/reviewers at *Screen* and Laura Mee and Johnny Walker and all at Cambridge Scholars Publishing. I am also very grateful to Philippa Brewster at I.B.Tauris for commissioning this work, to Anna Coatman for her help during the writing process, and to Maddy Hamey-Thomas for bringing it in to land.

Finally, I am hugely indebted to my friends and family for their love and support during the writing of this book. If I listed everybody who has lifted my spirits and provided a listening ear during the course of this work, these acknowledgements would be twice as long as they already are. However, special mention should be made of Phil Wheatley, and Claire Thompson and family, who looked after me during long trips to various archives; Vicky de Groot, who often held the fort whilst I made

Acknowledgments

these trips; and Merryll Wheatley, who put her life on hold every week to come and look after me and my family so I could get this book written. Kees de Groot has been at my side throughout; his belief in my ability to finish this work, and his willingness to help me to make this happen, have given me great strength. Books are tricky things to write, and no more so than when you simultaneously make three new human beings during the course of their production. Rudy Kornelis Wheatley de Groot, Dora Maisie Wheatley de Groot, and Arthur Leon Wheatley de Groot: you are my pride and joy. You have made this book infinitely more difficult to write, and life infinitely more fun to live. This book is dedicated to you.

Author's note

Chapter 2 of this book was originally published as '"Marvellous, awesome, true-to-life, epoch-making, a new dimension": Reconsidering the early history of colour television in Britain' in Laura Mee and Johnny Walker (eds) (2014) *Cinema, Television and History: New Approaches*, Newcastle-upon-Tyne: Cambridge Scholars Publishing. My sincere thanks to the editors for giving permission to reprint it here. It is published with the permission of Cambridge Scholars Publishing. Chapter 3 was originally published in 2013 in the *Journal of British Cinema and Television*, 10, 2: 257–75. Parts of Chapter 4 were originally included in my 2004 article 'The limits of television? Natural history programming and the transformation of public service broadcasting' *European Journal of Cultural Studies*, 7, 3: 325–39. Parts of Chapter 5 were originally included in my 2011 article, 'Beautiful images in spectacular clarity: Spectacular television, landscape programming and the question of (tele)visual pleasure' *Screen*, 52, 2: 233–48.

Introduction: What is spectacular television? What is (tele)visual pleasure?

'An unforgettable sight'
'A realm of fantasy, colour and excitement'
'Marvellous, awesome, true-to-life, epoch-making'
'A deliberate construction of events as both confronting and titillating'
'Gorgeous to behold: lump-in-throat, tear-in-eye beautiful'
'Beautiful images in spectacular clarity'
'Basically [...] you're just kind of looking at it going "Wow, look at that"'

So what is spectacular television and what is televisual pleasure? The descriptions of television programming at the top of this page all see viewers, programme makers, commissioners and reviewers capturing that sense of an enraptured gaze on television made to be appreciated visually, a gaze which lies at the heart of this work. The television I engage with here, made across television's 70-year history, is programming which is designed to be stared at, to be ogled, contemplated and scrutinised, to be gaped and gawked at. It is programming that absolutely conforms (at least in part) to the Oxford Dictionary's definitions of spectacle as both 'a visually striking performance or display' and 'an event or scene regarded in terms of its visual impact'[1] or to Geoff King's definition of spectacle as 'the production of images at which we might wish to stop and stare' (2000: 4). It is the image on the television which holds the viewer's gaze, and which, if only for a moment, can be appreciated outside of the drive of the narrative. Whilst this hyperbolic language of spectacle and visual impact might have been more readily used in the past to describe other media and entertainment forms, this book shows that there has been a simultaneous and ongoing history of televisual spectacle which has often been ignored in the rush to oversimplify and overdetermine television's medium specificity. This book thus contends that those who watch television and make television have frequently understood the medium as spectacular, as providing particularly visual forms of pleasure, but that these pleasures have been underplayed or underexplored by scholars of television. US television scholars such as John Thornton Caldwell (1995) and Mimi White, in her reconsideration of live television (2004), have brought our attention to the inadequacies of those cornerstones of television scholarship which, in establishing television's difference from cinema, have too quickly dismissed the medium's spectacular qualities. Typically, arguments about television that emphasise comparison with cinema position the medium as visually inefficient, as in Raymond Williams' proposal that:

> The domestic television set is in a number of ways an inefficient medium of visual broadcasting. Its inefficiency by comparison with the cinema is especially striking [...] Higher definition systems, and colour, have still only brought the domestic television set, as a machine, to the standard of a very inferior kind of cinema.
>
> (1974: 28)

However, whilst Williams' analysis of particular genres of programming suggested that in fact this visual inefficiency *didn't* lead to a lack of spectacle on television, John Ellis' later work on the medium as sound-led, lacking in visual detail, and appealing to the glance rather than the gaze did greater and longer-lasting damage when it came to the assessment and understanding of the (spectacularly) visual qualities of the television image. Ellis' work proposed that 'Broadcast TV cannot assume the same level of attention from its viewers that cinema can from its spectators' (1982: 115), given that the 'TV image is of a lower quality than the cinematic image in terms of its resolution of detail' (ibid.: 127). This, coupled with the competing demands placed on the distracted viewer's attention within a domestic setting, led to Ellis' proposal that:

> TV belongs to the everyday, to the normal backdrop of expectations and mundane pleasures [...] TV viewing is typically a casual experience rather than an intensive one. The consumption of TV is often described as 'relaxation' indicating a process that demands little concentrated attention, and is concerned with variety and diversion rather than enlightenment and excitement.
> (ibid. 160)

Whilst in fact spectacular television does indeed occasionally offer a form of relaxing 'diversion' (see Part II of this book), the sense that television (particularly in its spectacular forms) doesn't ever offer an intensive viewing experience, or enlightenment and excitement, must be challenged.

This analysis, then, sets itself apart from the extreme dualism of what Caldwell terms the 'mythology of essential media differences' (1995: 25) in which the forms and aesthetics of television and film must constantly be set at the opposing ends of two poles of viewer/spectator engagement and visual complexity/appeal. Specifically, this book seeks to engage directly with Caldwell's four correctives of 'glance theory' as they relate to this dismantling of essential media difference:

> Firstly, the viewer is not always, nor inherently, distracted. Secondly, if theorists would consider the similarities between television and film – rather than base universalizing assumptions on their 'inevitable' differences – glance and surrender theories would fall from their privileged pedestals. Third, psychoanalytic and feminizing extensions of glance theory tend to put critical analysis into essential and rigidly gendered straitjackets. Fourth and finally, even if viewers are inattentive, television works hard visually, not just through aural appeals, to attract the attention of the audience [...] Theorists should not jump to theoretical conclusions just because there is an ironing board in the room.
> (ibid.: 27)

Following Caldwell, this work proposes that the spectacular – an aesthetic category which has been so thoroughly and interestingly worked on and through in film studies – has been frequently ignored in television scholarship. Just as nominating intimacy and the everyday, the close up and the domestic, and an emphasis on dialogue over action as 'belonging to television' ignores whole swathes of film history which have been precisely formed around these foci and aesthetic qualities, so an assumption that spectacle and visual pleasure are solely affiliated with cinema misses an important aspect of television form as it has developed since the inception of the medium. Therefore, this book particularly focuses on the moments where television can be seen to be 'working hard visually [...] to attract the attention of the audience' (ibid.). Thus the concept of spectacular television might be related, in part, to what Caldwell describes as the stylistic 'exhibitionism' of televisuality (ibid.: 5).

However, whilst the analysis of television spectacle that proceeds from here is clearly indebted to Caldwell's ideas, there are some distinctions to make between his concept of televisuality and my mobilisation of the term 'television spectacle'. The first issue to deal with is the historicisation of these terms. Caldwell very clearly identifies the advent of 'televisual' (i.e. stylistically exhibitionist) television in the 1980s. Whilst he looks at some historical precursors to the programming of this moment, he argues that in the 1980s and into the 90s, '[US] television moved from a framework that approached broadcasting primarily as a form of word-based rhetoric and transmission [...] to a visually based mythology, framework, and aesthetic based on an extreme self-consciousness of style' (ibid.: 4). Further, he suggests that the development of technologies such as the video-assist and digital video effects helped to 'comprise an array of conditions [in this period], and a context that allowed for exhibitionism' (ibid.: 7). Thus Caldwell's discussion of stylistically exhibitionist television is situated in relation to a particular moment of broadcasting and a particular set of production contexts. Furthermore, whilst challenging Ellis' glance theory, Caldwell wonders whether the concept is based on the viewing of 'primitive shows produced in the early, formative years of the medium' (ibid.: 25). In doing so, then, he reaffirms the assumption that the concept of televisuality is historically bounded, relating to a particular moment in (US) TV history, and that non-televisual television came before.

In contrast, this book demonstrates that television schedules have incorporated the spectacular throughout the medium's history. Whilst individual chapters look at particular moments when spectacular forms and programming genres have been brought to the fore (as in the coming of colour television in the late 1960s, or in the shift to HD filming and the migration of television sets from the hearth to the wall in the first part of the twenty-first century), this book argues more broadly that television has *always* had moments, programmes and genres which can be identified as spectacular, and has always incorporated visual pleasure into its schedules. However, whilst this book seeks to historicise our understanding of the spectacular on television, it does also acknowledge that, as John Corner puts it, 'visual treats are one of the key lures in a multichannel age' (1999: 35) and engages with Karen Lury's suggestion that a greater emphasis on visual pleasure is brought about by contemporary digital technologies (2005: 14). Three sets of changes – greater choice and competition in broadcasting, changes in image quality and clarity and changes in the screens on which television is watched – are explored in this book as context for the provision of spectacle and visual pleasure on television, and each of these sets of changes is related both to the contemporary moment in television broadcasting and to television's broader history. Whilst the case studies discussed here are drawn largely from British television history, television's interest in spectacle is not particular to British programmes or genres, for as Mimi White argues in relation to both US and European programming 'Television's ability to *show* things, and its interest therein, has been unduly muted by theories of the glance [...] Yet television has demonstrated an abiding interest in visual spectacle for its own sake, in the televisual attraction' (2004: 85). Whilst White discusses the 'television of attractions', reformulating Tom Gunning's classic 'cinema of attractions' theory (1986) in order to think about 'television's proclivities for spatial display' (White, 2004: 76), the programming she is discussing (made in a variety of US and European contexts) can also be considered as spectacular television, which, according to the definitions above, forms a 'visually striking performance or display', 'an event or scene regarded in terms of its visual impact', and 'the production of images at which we might wish to stop and stare'.

Another fundamental difference between this analysis of television spectacle and Caldwell's discussion of televisuality relates to the question of viewer demographics. An address to an educated, affluent, 'yuppy' viewer is central to Caldwell's reading of televisuality; he sees the increase in stylistically exhibitionist programming as a key way in which this lucrative audience was targeted by the networks, though he is also interested in what he calls 'trash spectaculars' – e.g. *Rock and Rollergames* (KTLA-Channel 5, 1989), *American Gladiators* (MGM, 1989–96), *Knights and Warriors* (Welk Entertainment, 1992) – which clearly don't fit with this delineation of the audience for televisual television. On the other hand, however, this book proposes that television spectacle can be located across genres, programmes and channels appealing to diverse audiences, and is not solely produced to appeal to a particular viewer demographic. The book argues that certain examples of spectacular television are formed around particularly classed taste codes: analyses of wildlife programming and television landscapes engage with the ways in which certain forms of television spectacle are designed to be aspirational and to draw on established upper-middle-class taste cultures; several of the 'dramas of sex' discussed here fall into the 'quality television' category which has been traditionally understood as appealing precisely to those 'yuppy' viewers identified by Caldwell. Alternatively though, other programming discussed in this study has much broader appeal to a popular, 'primetime' audience, often drawing on more working-class traditions of spectacular presentation; this is evident, for example, in the forms of reality programming and popular documentary series discussed herein (the reality-holiday series, or the fascinoma documentaries, with their roots in the carnival sideshow), and also in the relationship between the 'shiny floor shows' (the studio-based entertainment programming discussed below) and the presentational traditions of variety and vaudeville performance. Indeed, it is clear that given that British television drew on theatrical vaudeville and music hall performance for many of its stars, entertainment forms and production personnel from its inception, and that this relationship was particularly stressed in the demonstration and exhibition of television in the mid-twentieth century, spectacle on television is in part grounded in the specific presentation and performance traditions of British working-class entertainment cultures.

Before moving on from this comparative analysis of my approach to television spectacle and Caldwell's televisuality concept, the question of the cinematic must be addressed briefly. In Caldwell's analysis, he argues that one of the ways in which television became stylistically exhibitionist was in its move towards a 'film look': 'Exhibitionist television in the 1980s meant more than shooting on film [...] Rather, cinematic values brought to television spectacle, high-production values, and feature-style cinematography' (1995: 12). In the wake of this argument, many 'high-end' television dramas have been labelled 'cinematic' within television scholarship, a critical move which, as Deborah L. Jaramillo has argued, limits our full understanding of the development of television aesthetics:

> 'Cinematic' should be a contentious word in the field of television studies. It should raise the eyebrows of anyone who thinks and writes about television; instead it has become commonplace for scholars and popular critics to use the term as shorthand when discussing [a] complex visual and aural style in scripted series [...] *'Cinematic' connotes artistry mixed with a sense of grandeur* [...] It is an inherently positive, even boastful word that many people rally around and ascribe to the best of the best on TV.
>
> (2013: 67 (my emphasis))

In exploring a variety of different forms of spectacle on television, this book rejects the idea, as Jaramillo does, that spectacular television is inherently cinematic. It is proposed here rather that some of the critical language developed around questions of spectacle and visual pleasure almost exclusively associated with film and Film Studies, particularly that which describes the intensity of the relationship between viewer and audio-visual text, might be mobilised to explore a whole series of programmes, genres and modes of television across its history.[2] This does not constitute an argument that spectacular television is cinematic, but rather that the aesthetics of spectacle more usually associated with film might equally be associated with certain forms of television: that film and television are both, at times, spectacular, rather than that television is cinematic because of its presentation of spectacle and its various associated visual pleasures.

The spectacular in audio-visual culture: Debates and problems

As argued above, theories and histories of the spectacular are well developed in relation to the analysis of film. Questions about the place of spectacle within narrative cinema, and what that spectacle might be distracting film audiences from or drawing them towards, have flourished in analyses of specific genres (the musical, or action cinema, for example) and particularly in relation to Hollywood cinema. Much of this debate has rested on the question of a potential imbalance between narrative and spectacle, where spectacle threatens to overwhelm the spectators, distracting them from either a proper engagement with narrative or from a critical engagement with questions of ideology. As Peter Kramer has documented, 'the displacement of narrative by spectacle has [...] been the subject of critical debate [since] the 1950s' (1998: 303). This debate is seen, for example, in Martin Rubin's discussion of the spectacle of Busby Berkeley's musical numbers in which he argues that

> Of crucial importance to the creation of Berkleyesque spectacle is a sense of gratuitousness, of uselessness, of extravagance, of rampant excess, of over-indulgence, of flaunting, of conspicuous consumption, of display for the sake of display, of elements calling attention to themselves rather than serving a higher, all-encompassing concept of narrative.
>
> (1993: 41)

It is also seen in Justin Wyatt's exploration of high-concept cinema (1994: 25) and in Norman King's seminal analysis of the films of Abel Gance, in which he argues that 'we could define filmic spectacle as that which arrests the spectator's look at the surface of the frame as opposed to what is within the frame' (1984: 201). Lea Jacobs and Richard de Cordova's early work on cinematic spectacle proposed that 'the shift from narrative to spectacle is predicated on a slippage between the event as fiction and the image as construction. In spectacle, the figuration of the visual field takes up the function conventionally held by narrative action' (1982: 301). In relation to television, this debate about the balance between narrative and spectacle abides, both in this work and in the broader discourse surrounding particular programmes. For example, Chapter 4 discusses the combination of moments of heightened visual pleasure with a focus on the 'education' of the viewer in natural history series, and, subsequently, the need for broadcasters such as the BBC to negotiate their public service remit to 'inform, educate, and entertain' through this balancing of the provision of visual spectacle and a more didactic narrative focus. This balance is both discussed in this book's analysis of key natural history programmes and is also shot through the discussions of these programmes offered by programme and policy makers, reviewers and viewers alike. More broadly though, the facts of television broadcast, with its seemingly 'endless' flow of programming, mean that it can often feel like the spectacular 'erupts' into the ongoing broadcast text, sometimes as a potentially disruptive force; this idea is explored at length in the discussion of the representation of death on television in Chapter 6, for example, and in the analysis of television eroticism in Chapter 7. Here spectacle might perhaps be understood as interrupting the broader 'flow' of television as well as the narrative of individual programmes.

In film history, spectacle has also been understood as a response to increased competition: Geoff King proposes that 'An emphasis on spectacle formed a central part of a post-war strategy aimed at tempting lost audiences back to the cinema in the face of demographic changes and the development of television and other domestic leisure activities' (2000: 1). This argument is also corroborated in Peter Lev's history of American cinema in the 1950s, in which he suggests that a turn towards spectacle was all about showing what film could do better than television (particularly in a shift towards colour filming in this period) (2003: 107). As discussed above, television has also turned towards the spectacular at times of increased competition and during periods of technological change, perhaps no more so than in the contemporary television landscape where programmes broadcast on TV must compete with streaming/downloading services (Netflix, Amazon Prime, Hulu, Now TV), alongside a proliferation of channels in the multichannel environment and other forms of media consumption more broadly. However, the production of spectacle can also be seen as a competitive strategy in the history of British television in the late 1960s (in relation to the inception of BBC2, and the introduction of colour), or earlier in the marketing strategies of the BBC and the ITV companies at the outset of competition in the mid- to late 1950s. In essence then, television broadcasters have been continually responding to competition (from other forms of leisure/entertainment, from the distractions of everyday life and, internally, from other channels on the dial or the electronic programme guide); this means that the recourse to the spectacular, the visually arresting, the 'images at which we might wish to stop and stare', has been one of a number of strategies used to distinguish programming throughout the history of the medium.

Film theory has also explored the politics of spectacle, taking its lead from the writing of Guy Debord (1995) in the late 1960s, and questions of how ideology and power operate in relation to this. This work figures the spectator as 'locked into' spectacle as a totalitarian visual system which denies us free will or intellectual autonomy, which takes hold of the viewer-spectator and won't let them go. Hal Foster, for example, argues that 'We become locked in its logic because spectacle both affects the loss of the real and provides us with the fetishistic images necessary to deny or assuage this loss' (1985: 83). Similarly, Leger Grindon proposes that 'The spectacle is currently a suspect element whose ideological operations infect the film viewer unaware' (1994: 35), drawing on Thomas Elsaesser's argument that 'the cinema displaces ideological coherence into spectacle' (1986: 26) and Dudley Andrew's delineation of filmic spectacle as designed to entice the viewer into becoming a 'plaything of the movies' (1984: 122). This politically radical position demonstrates a deep distrust of spectacle: it proposes that spectacle distracts us temporarily not only from the progression of narrative in a film, but also, potentially, from engaging critically or thoughtfully with the world around us.

In television theory, this position is taken up in Douglas Kellner's analysis of the 'triumph' of spectacle on TV (2005) and, in a more sustained way, in Neil Postman's work in which he reflects on television's role in what he calls the 'peek-a-boo world [...] a world without much coherence or sense; a world that does not ask us, indeed, does not permit us to do anything' (1985: 78–9). Discussing American television, which he calls 'a beautiful spectacle, a visual delight' (ibid.: 88), Postman argues that 'It is in the nature of the medium that it must suppress the content of ideas in order to accommodate the requirements of visual interest' (ibid.: 94). In this line of argument, and, for example, in Whannel (1984) and Tomlinson's (1996) discussion of sports broadcasting, or Stratton and Ang's (1994) analysis of the reality/fly-on-the-wall documentary series, spectacle overwhelms both viewer and medium, replacing what is potentially authentic or important with the distracting and the titillating, blurring the lines between fiction and reality, and, fundamentally, replacing discourse, dialogue and debate with the image. This debate is what John Corner is referring to when he argues that 'It is the properties of the image which fascinate and attract, which become the focus for debate about sexually explicit and violent content, which are featured in most discussion of advertising and of the displacement of political substance by "presentation"' (1999: 24). It also sits at odds with distraction theory because in these works of television theory and analysis, the television image is visually powerful, arresting even, as opposed to lacking in visual appeal. Whilst we might challenge the ideas of Postman and others on the grounds that they in fact overstate the spectacular nature of the medium and propose a rather singular and simplistic relationship with television spectacle which fails to take into account the diverse ways we might engage with television, this work provides an important early critical context to this book.

The other significant piece of theoretical writing that precedes and contextualises this reading of spectacular television is John Fiske's earlier analysis of television form. Fiske's work, perhaps somewhat overshadowed by the pervasiveness of glance theory, constitutes an early analysis of the *pleasures* of television spectacle, although in some ways it is not in disagreement with the ideas discussed above. For example, he proposes that:

> The spectacular involves an exaggeration of the pleasures of looking. It exaggerates the visible, magnifies and foregrounds the surface appearance, and refuses meaning or depth. When the object is pure spectacle it works only on the physical senses, the body of the spectator, not in the construction of a subject. Spectacle liberates from subjectivity.
>
> (1987: 243)

Here, the viewer is again presumed to be drawn to the surface of the image, away from a critically engaged viewing position, but rather than focusing on the negative aspects of this relationship with an all-consuming spectacular image on television, Fiske focuses on this as a fully embodied position of *pleasure*. He acknowledges, for example, the *jouissance* produced by exaggerated images of high emotion in the soap opera (ibid.: 229), and, furthermore, in his exploration of a Bakhtinian notion of the carnivalesque (Bakhtin, 1968), Fiske critically engages with the transgressive potential of television spectacle to provide textual 'eruptions' of excess. Ultimately, however, Fiske qualifies his engagement with the spectacular on television in the following way:

> [The] spectacle of television is not the grotesquerie of carnival (though *Rock n Wrestling* approaches close to it), nor is it the panoramic spectacle of cinema, it is a street-scaled spectacle dimensioned around and on the human body. It engages not the power-bearing, subjugating look of Mulvey's cinema, but a look of equals, for the looker participates in the spectacle, the looker can be looked at.
>
> (1987: 264)

Whilst we can take issue with his dismissal of both the spectacle of the grotesque and the spectacle of the panorama as key visual pleasures of television, Fiske's work initiates thinking about the nature of televisual spectacle, and its power to engage and enthral the viewer, which this book picks up.

Categorising spectacular television: Genre, movement or mode?

Television is, of course, a heterogeneous medium, never one thing or another, but rather a patchwork of different forms and genres. It thus follows that not all television is spectacular, and indeed that not all spectacular television is spectacular all of the time or in the same ways. Whilst we ought to pay more attention to the spectacular and to visual pleasure on television, it may not yet be clear whether in using the term 'spectacular television' we are referring to a particular genre of programming, or a historical movement or movements, or a broader mode or style of programming. The answer to this is, in fact, that spectacular television is and has been all three of these things.

First, in relation to the early history television in the UK, 'the Spectacular' was a term used to generically define 'special' light entertainment shows throughout the twentieth century, as well as referring more broadly to the genre of variety programming which had its heyday in the 1950s and 60s. In the UK, one-off variety specials (e.g. *Scintillation!* BBC, tx. 14/7/51), long-running variety programmes (e.g. *Val Parnell's Saturday Spectacular*, later just *Saturday Spectacular*, ATV, 1956–61), ice shows (e.g. *Carnival on*

Ice, ITV, 1956; *Holiday on Ice 1960*, BBC, tx. 24/1/60), circus shows (e.g. *Billy Smart's Circus*, BBC, 1947–78; or the *Moscow State Circus*, ITV, 1959–61) and 'Spectaculars' from the US (e.g. *Britain Presents America's Best*, ITV, tx. 7/12/57, billed in the *TV Times* as 'the first great American Spectacular to be shown in Britain'; and the *Jack Benny Spectacular*, broadcast in the UK by the BBC in 1959 and 1960) were popular throughout this period. Frequently situated in the middle of primetime on Saturday evenings, and directed towards a family audience, these variety programmes, described by Raymond Williams as 'tinsel and plush spectaculars [...] glimpses of "high life"', and expensively furnished display' (1974: 65), drew directly on vaudeville and music hall traditions of entertainment and offered a version of the spectacular which could be performed on stage or in an arena in front of a diegetic audience. This genre of programming was particularly important in both the US[3] and the UK during the launch of colour television, where it was seen to offer its viewers a 'dazzling' spectacle of light, sound and colour, and to really show off the capabilities of this new technology. Susan Murray has argued that the 'Spectacular' was produced as the 'industry desired its performers to utilize many stage techniques – both legitimate and vaudeville styles – in order to flaunt the visuality of the new medium' (2005: 51); here, then, the emphasis on what can be seen at the level of performance renders the spectacular a genre of programming inherently invested in visual spectacle.

There is more work to be done on the history of the spectacular as genre (in fact there is relatively little discussion of this programming in this book), particularly in understanding it as a precursor for the contemporary 'shiny floor' show, popular Saturday night TV light entertainment that ranges from the big talent shows (*X Factor*, ITV1, 2004–; *Britain's Got Talent*, ITV1, 2007–; *The Voice*, BBC1, 2012–), to competitions (*Strictly Come Dancing*, BBC1, 2004–; *The Cube*, ITV1, 2009–), celebrity-focused light entertainment shows (*Ant and Dec's Saturday Night Takeaway*, ITV1, 2003–; *The Graham Norton Show*, BBC1, 2007–), dating programmes (*Take Me Out*, ITV1, 2007–), and the para-sports programmes that Caldwell terms 'trash spectaculars' (1995: 3) (e.g. *Gladiators*, ITV1, 1992–2000; *Ninja Warrior UK*, BBC1, 2015–). Whilst this list of programmes is diverse in terms of programme formats, structures and performance styles, these shows are united aesthetically by a sense of gloss, by bright lights, glitter and shine, and by the transformation of the television studio or arena into spectacular space. There is a grandiosity to these programmes, and a sense of glamour (particularly in relation to the styling of female presenters and participants), which relates back directly to Williams' 'tinsel and plush spectaculars' and his 'glimpses of "high life", and expensively furnished display' (1974: 65). An analysis of this programming and the historical trajectory to be traced from variety and music hall traditions to these shows is thus still to be done.

However, just as José Arroyo has argued in relation to contemporary action cinema that the spectacular films he is interested in are linked less by genre than to a 'similar production profile (high-budget, lavishly produced potential blockbusters that anchor a studio's seasonal release schedule)' (2000: vii), so we might think beyond generic boundaries when considering television spectacle. Outside of thinking about the spectacular as a well-defined genre of programming with a long history, it is possible to identify *movements* of spectacle that run across genres, particular moments in production where spectacle and visual pleasure are brought to the fore, as with the introduction of colour television in the UK in 1967, or with the gradual take up of HD technologies from the mid-2000s onwards. As argued in Chapter 5, an increased emphasis on the spectacle of landscape cinematography was seen in such diverse genres as the history/natural history documentary series, the cookery programme, the heritage drama and children's television, marking a broader movement of the spectacular across television as a whole during this period. Beyond these historically specific movements of heightened visual pleasure on television, however, spectacular television might also be seen as a mode or style within particular programmes, and thus spectacular television might more loosely be understood as the programming that asks us to *look*, to observe closely or to contemplate. It is, fundamentally then, that programming in which our attention is drawn to the fact of looking.

Mimi White's discussion of such programming as a 'television of attractions' is illuminating:

> Television's ability to show things, and its interest therein, has been unduly muted by theories of the glance, of the medium's reliance on sound over image to convey meaning and regulate

> spectatorship, and of liveness as a fundamentally temporal category. Yet television has demonstrated an abiding interest in visual spectacle for its own sake, in the televisual attraction.
>
> (2004: 85)

White particularly engages with those programmes where often un-narrated images implore viewers to look: she analyses an episode of the news magazine/documentary series *See It Now* (CBS, 1951) in which Edward R. Murrow presented simultaneous views of the Atlantic and the Pacific Oceans, arguing that 'Here, space functions as spectacle, a visual attraction transmitted in the present tense of live television' (ibid.: 84). Additionally, White analyses traffic and weather programmes from Spain and Austria, and the Home Shopping Club in the US, in which 'relatively static images are shown for a considerable duration, often with little or no variation' (ibid. 85). In fact, the programmes White discusses in her analysis of the 'television of attractions' can be seen formally as precursors to a movement of distinctly unspectacular spectacular television: 'slow TV'. The slow TV concept, the 'slow' designation an adaptation of Andy Warhol's 'slow movie',[4] refers to a European genre of programming which began with German satellite channel Bahn TV's daily broadcast of driver's eye view films of German rail lines under the title *Bahn TV In Fahrt* (2003–8). Following this, slow TV became an unlikely hit in Norway in 2009

> with a seven-hour film about a train journey, followed by a 12-hour knitting marathon and the live broadcast of a five-day boat trip [from Bergen to Kirkenes] which had thousands of people lining the route and was watched by more than half of the Norwegian population.
>
> (Plunkett, 2015)

Norway's slow TV movement also produced *Nasjonal vedkveldeight* (NRK, 15/2/13), featuring eight uninterrupted hours of footage of a burning fireplace.

After the huge popular success of Norwegian slow TV, the BBC commissioned a season of programmes for its 'grown up' channel BBC4, broadcast under the umbrella title 'BBC Four Goes Slow' during the first full week of May in 2015. Cassian Harrison, the editor of BBC4, said of this programming:

> We are so used to the conventional grammar of television in which everything gets faster and faster, we thought it would be interesting to make something that wasn't continually shouting at you and coming up with the next climactic moment [...] [These programmes allow viewers to] enjoy an experience in more detail and intimacy than television normally allows, and sit back and draw your own conclusions.
>
> (ibid.)

Dawn Chorus: The Sounds of Spring, broadcast in the early hours of Tuesday 5 May at the beginning of this week-long season in 2015 and offering 'The birdsong of sunrise in all its uninterrupted glory, free from the voiceover and music of traditional television,'[5] captioned a montage of landscape images and shots of birds singing with very occasional identifying labels, and focused the viewer's attention not only on what could be seen but also on what could be heard. A three-part series titled *Handmade* brought together episodes without narration, showing the production of a glass jug, a steel knife and a Windsor chair, and emphasised the skills involved in these traditional crafts as well as the challenge of sitting and watching television without someone on or off screen instructing you to look here or there, at this or that. *All Aboard! The Canal Trip* (tx. 5/6/15), the centrepiece of the season, was a two-hour, uninterrupted journey along the Kennet and Avon Canal shot from the front of a narrowboat, which again had no narration but incorporated CGI intertitles with facts about the canal that were positioned to look as if they were written on the water or on the boats passed by the camera. According to Catherine Johnson, 'These moments convey the kind of information that would typically be provided through voice over or interview [but] far from a distraction, historical and factual information here becomes part of the landscape and the journey' (2015).

Slow television thus offers 'unspectacular' spectacle in that it features none of the 'whistles and bells' of spectacular television as described above, no dramatic lighting or particularly obtrusive cinematography, but rather offers precisely what Martin Lefebvre calls 'space freed from eventhood' (2006a: 22) in his analysis of landscape in cinema, showing the viewers spaces, places and activities at which they are simply invited to look. Whilst the journey at the heart of *All Aboard!* might be seen as an event in and of itself, what we are asked to look at along the way is distinctly uneventful;

thus this programming fits precisely with White's delineation of the 'television of attractions', understood here as a form of unspectacular spectacle. Unusually, this programming is 'pure' or 'all spectacle' in this sense, whereas the majority of the television discussed in the remainder of this book is better understood as providing *moments* of spectacle, moments, sometimes no longer than a shot, in which we are asked to *look*, to appreciate, to become absorbed in the images on screen.

This book does not attempt to provide an exhaustive or encyclopedic history of spectacular television or televisual pleasure (by which I mean television's specific visual pleasures, rather than using Caldwell's repurposing of the word 'televisual'). Rather, it offers a series of case studies or a set of 'ways in' to thinking about the spectacular and visual pleasure in relation to television that are designed to provoke further thought and inspire further research, and which speak to the genres and moments noted above. The book takes three turns in exploring spectacular television. Part I of the book, 'Spectacular Histories, Spectacular Technologies', looks at two historical case studies in which television figures as a spectacular technology. In this section, Chapter 1, 'Television comes to town: The spectacle of television at the mid-twentieth-century exhibition and beyond', looks at how television itself was presented as spectacle of modernity at large-scale public exhibitions such as Radiolympia (later the National Radio Show, 1926–64), the Ideal Home Exhibition (1908–) and the Festival of Britain (1951), exploring what John Hartley describes as television's position as a 'popular attraction associated with crowds, spectacle, urban activity, technology and modernization' and its centrality in the 'mid-century passion for spectacular national self-aggrandizement [...] and for the public display of industrial innovation as the aesthetic of the age' (1999: 75). Chapter 2, 'Spectacular colour? Reconsidering the launch of colour television in Britain', focuses in on the history of the launch of colour in the UK and the negotiation that took place in this period between the push to produce 'spectacles of colour' and a pull towards a more subdued, more naturalistic presentation of the capabilities of colour TV.

In Part II of the book, 'Spectacular Landscapes and the Natural World: Exploring Beautiful Television', focuses on a discussion of what John Corner calls 'realist visual pleasures' (1999: 94), 'looking at pictures of things which are themselves pleasurable' (ibid.). Chapter 3, 'At home on safari: Colonial spectacle, domestic space and 1950s television', offers a historical case study of spectacular natural history television and explores the ways in which the people, animals and landscape of Kenya are presented as spectacle in the work of the colonial programme makers Armand and Michaela Denis. As such, this chapter offers an examination of a very specific moment in the history of British broadcasting in order to understand the ideological significance of televisual spectacle, and the looking relations at play in rendering a particular set of images spectacular. This case study suggests that television itself might be understood as a 'colonising' apparatus, and looks at the ways in which metaphors of the colonial have become rooted in our understanding of television as a medium of spectacle. Chapter 4, 'Visual pleasure, natural history television and televisual beauty', revisits some earlier work on natural history television, public service broadcasting and visual pleasure as a springboard for the consideration of the idea of beauty as it relates to television. Here, and in Chapter 5's exploration of television landscapes, this concept is interrogated, and the ways in which beauty has been identified as one of television's key pleasures is considered. This chapter in particular attends to the ways in which beauty (beautiful programmes, beautiful moments in programmes) has been defined in television policy and production, as well as in the discourses surrounding television programming (in reviews, and in viewer evaluation of programming). Beauty emerges as a (largely unspoken) criterion in the assessment of 'quality' television, and this might be related to the way in which programming can fulfil one of Katz et al.'s basic functions of television: 'the need for emotional and aesthetic experience […] the desire to see beautiful things' (1973: 173). The final chapter of this section, 'Television's landscapes, (tele)visual pleasure and the imagined elsewhere', explores televisual landscape in two kinds of programming: firstly, the landscape-focused documentary series such as *Bird's Eye View* (BBC, 1969–71), *A Picture of Britain* (BBC, 2005), *Coast* (The Open University/BBC, 2005–), *Britain's Favourite View* (ITV, 2007) and the *Wainwright Walks* series (Skyworks for BBC4, 2007–9), and secondly, the holiday programme, in which landscape is a key signifier of an imagined elsewhere presented to the viewer-tourist. Across both case studies, the idea of visual pleasure in non-fictional television is explored and related to a contemplative mode of viewing more traditionally associated with the spectacular in other media and at odds with both theories of the distracted viewer identified by early theorists of television (e.g. Ellis, 1982) and counter-theories of 'sit forward' viewer engagement or enthralment (developed particularly in relation to describing the viewing of recent 'quality drama' and in relation to Caldwell's notion of 'televisuality' (1995)). This is programming that denies John Corner's suggestion that the 'wide-angle landscape shot' is absent or lacks detail on television (1999: 26), but rather relates to the specific historical contexts (the shift to HD filming and the migration of television sets from the hearth to the wall in the first part of the twenty-first century) which saw precisely this kind of filmmaking all over British television.

Part III, 'Spectacular Bodies and (Tele)visual Pleasure', focuses on the spectacle of the body on television in two very different contexts. Firstly, in Chapter 6, 'Fascinating bodies: Looking inside television's somatic spectacle', consideration is offered, for example, of the rush of a CGI roller-coaster ride through the human digestive system in 'human body' documentary series, the televisual allure of fascinoma (rare medical cases) and the presentation of bodies in popular factual entertainment shows. Further, this chapter also explores the limits of intense viewing pleasure in looking closely at bodies when we are confronted with images of death and abjection on the small screen. In the concluding chapter, 'The erotics of television', we turn to a consideration of the powerful spectacle that desiring and desirable bodies provide, and in doing so re-evaluate the applicability of cornerstones of screen theory to television, particularly Laura Mulvey's 'Visual Pleasure and Narrative Cinema' (1975). This part of the book proposes that just as Linda Williams was able to identify the 'body genres' of cinema (the horror film, melodrama and pornography) in her exploration of the 'spectacle of a body caught in the grip of sensation or emotion' (1991: 4), so it is possible to define the body genres of television.

Finally, the conclusion of *Spectacular Television* presents an analysis of *The Opening Ceremony of the London 2012 Olympic Games: Isles of Wonder* (BBC, 2012), in order to draw together the various strands of the book's analysis of spectacle and visual pleasure. The conclusion shows that this high-profile example of spectacular television offers a kind of catalogue of television's visual pleasures as they have been explored here. I also use this concluding analysis to reiterate the fact that although this is clearly an example of special 'event' television which is shot through with multiple forms of television spectacle, in fact spectacle is inherent within and across a myriad of televisual genres and throughout the medium's history; it is not reserved for 'special' or 'ceremonial' programming, but rather is part of the day-to-day fabric of television broadcasting.

This introduction to the book must end with a brief reflection on the methods used here in order to interrogate spectacular television and televisual pleasure. As I have argued elsewhere, a multi-methodological approach is needed in order to produce a fully rounded account of television and television history (Wheatley 2007b: 8), drawing on Ann Gray's proposal that we 'conceptualise historical research methodologies as a kind of contingent mosaic, in which television historians draw together different strands of the production/text/viewer triumvirate according to the particular needs of the project' (ibid.). In this book, textual analysis is combined with research into television production and the TV industry, an interest in how television is marketed and an engagement with viewer research. The latter, undertaken via a combination of face-to-face interviews and observations, follow-up questionnaires and online surveys circulated using a snowballing technique via social media,[6] offers in particular a snapshot of the ways in which the 'ordinary viewer' engages with and enjoys spectacular television. This gives an indication of understandings of key terms in this study – spectacle, visual pleasure, the beautiful, enthralment – in wider (non-academic) circulation. More broadly, it is clear from the viewer research in this study, and from the analysis of the discourses that circulate about spectacle within the television industry, that whilst an emphasis on visual pleasure may not be fully or firmly situated within the scholarly analysis of television thus far, viewers and programme makers alike continue to understand the spectacular to be one of the key pleasures of television.

Part I

Spectacular Histories, Spectacular Technologies

1

Television comes to town: The spectacle of television at the mid-twentieth-century exhibition and beyond

When a group of journalists visited Fleet Street on 6 October 1936 as guests of the Marconiphone Company, several weeks before the BBC began its regular broadcasts of television, they were treated to the following demonstration:

> On a screen that was about ten inches by eight inches we saw in a sepia tone a moving picture which was quite as clear in definition as any home kinema [...] First we saw the Alexandra Park grounds with people walking about little knowing that they were being televised [...] We toured by television an exhibition that was to open the next day in the palace. We saw Leonard Henry doing comic Shakespearean acts – in a gas mask! This was close-up and had very definite entertainment value. We saw a mannequin parade, dancing, a band, [and] announcers.
>
> (May, 1936)

Here we see television in Britain as spectacular from its outset. This pre-broadcast demonstration, combining an outside broadcast which transformed Alexandra Park and its unknowing visitors into pure spectacle (an image to be looked at without a clear narrative function) alongside various variety acts delivering the kind of performances (vaudeville humour, dance, band music and a fashion show) which would be collectively grouped under the generic title of the 'spectacular' show well into the 1960s at the BBC, presents television as a spectacular medium, dwelling on what it can show rather than what it can say. That this demonstration included a tour of an exhibition which was to take place in Alexandra Palace[1] is also significant, given that the mid-century exhibition was to become such an important site in the history of British television. The exhibition was a site for the demonstration of television, a place in which (often spectacular) television would be produced, and a space where the significance and placement of television in the home would be deliberated. This chapter is thus interested in the ways in which television was presented as a medium of spectacle at the exhibition and was figured as a significant part of the spectacle of modernity in mid-twentieth-century Britain.

For many people in Britain, events such as Radiolympia (later the National Radio Show, 1926–64), the Ideal Home Exhibition (later the Ideal Home Show, 1908–) or the Festival of Britain (1951) would be the place where they would first encounter television, or first encounter new developments in television technology. For example, whilst Brian Winston asserts that 'barely 2,000 sets' were sold in the first year of television broadcast in the UK (1998: 112), it was reported in the *Ideal Home* magazine in November 1937 that a further 5,000 people had seen television in the past year at 'Radiolympia, or the Science Museum at Kensington, or in a shop or a friend's home' (Anon., 1937) making the exhibition a key location for first encounters with television. The mid-century exhibition is thus an important, and largely overlooked, site in the history of British television, as it is in the US and beyond,[2] in the narrative about the introduction of the medium to the public. This was a series of exhibitions that combined set manufacturers demonstrating their latest models and the ways in which they should be placed into a series of 'ideal homes', with exhibits made by or for the BBC (and later the ITV companies) in order to reflect on their technical, artistic and public service achievements. This chapter reveals that the large-scale public exhibition played a vital role in the early negotiation of the ontological status of television: what it was or is, what it would be, what it offered for a particular group or groups of viewers. This is why the history of television at the exhibition is of broader interest to cultural and media historians. Just as Anna McCarthy's work on television in a variety of public spaces has explored the medium's 'ability to collapse distinctions between public and private space' (2001: 32), illuminating a historically situated notion of medium specificity, so this chapter examines the ways in which often competing ideas about television's medium specificity in the mid-twentieth century were exposed at the exhibition. It also takes its lead from William Boddy's work on the launching of television and other media in the US (2004), in that particular attention is paid to what promotional strategies can tell us about industrial, institutional and popular understandings of what a medium is and what it can do. Whereas McCarthy's exploration of 'ambient television' sought to understand the quotidian geography of television in public spaces, this chapter historicises what might be termed the 'spectacular geography' of public television, and, at its conclusion, offers some consideration of more recent events and spaces where television has been placed at the boundaries of broadcast, exhibition and live performance. In attending to the history of television at the mid-century exhibition, the origins of the conception of television as a medium of spectacle are revealed. An account of the appearance of television at the Festival of Britain, offering a case study of the ways in which television was figured at the mid-twentieth century exhibition, is thus an appropriate starting point.

The Festival of Britain

The Festival of Britain, encompassing seven large government-sponsored exhibitions and the Battersea Pleasure Gardens, took place around the UK between May and September in 1951. It was conceived of as a wide-ranging celebration of Britain's contribution to science, technology, industrial design, architecture and the arts on the centenary of the Great Exhibition of 1851, and as a way to stimulate (or simulate) national recovery in the aftermath of World War II. The central exhibition at London's South Bank offered a sense of celebration and liberation, a new-found freedom centring on a nation moving out of a period of deprivation and towards one of increased leisure and consumer opportunities. This part of the exhibition featured the futuristic architectural designs of the Skylon structure and the Dome of Discovery, transforming the South Bank into a site of wonder. In all exhibits, histories of past accomplishments were interwoven with narratives of a promising future for the nation. Discussing the appearance of the festival in family photograph albums, Harriet Atkinson describes 'images of emancipation, of a populace reconnecting with itself after severe and extended privation, of people, many still in demob clothes […] enjoying their first day out in years' (2012: 3).

The bringing together of science, commerce and the arts in the South Bank was stated and restated throughout the Festival and in the rhetoric that surrounded it. For example, in his speech to close the ceremony, the Archbishop of Canterbury said

> The Festival, like the Dome of Discovery itself, was marked by imagination and ingenuity; a fearless gaze both into the vastness of space and into the minute details of every art and science; a pride in what Britain has achieved in things spiritual, cultural, scientific and industrial; a sense of what honest work and cooperation can do.
> (Fisher, 1951)

This rhetoric, in which science, arts and commerce are combined to spectacular effect, was also to be found in the promotion of television at the Festival, given that, as Sarah Street has argued, 'The celebration of scientific achievement […] was presented as part of an exciting *longue durée* from Newton and Darwin to televisions, jet engines and radar' (2012: 84). An advertisement for Cossor Electronics (Fig. 1.1) in the festival catalogue claimed television as a prime example of British scientific achievement: 'Radio and Television […] are typical of the vast resources of Cossor experience and ingenuity, a great British national asset today – and in the future development of this electronic era' (Cox, 1951: xix). However, in an Ekco ad placed several pages ahead (Fig. 1.2), television is lauded as a medium of the arts. This advertisement, bordered by a gilt picture frame, shows a painted image of a diva in lavish gold ball gown, as if lit from below. The accompanying text, reworking Shelley's 'To the Skylark' (1820) reads:

> All that ever was joyous and clear and fresh – thy music does surpass: To present the full beauty of the human voice, the orchestra, the solo instrument: this is the proud achievement of one of the really great names in Radio and Television […] Ekco.
> (Cox, 1951: xxviii)

Fig. 1.1 Cossor Electronics advertisement from the Festival of Britain catalogue (1951)

These two advertisements thus succinctly encapsulate television's placement at the centre of the Festival of Britain's dual aims to celebrate the scientific and cultural achievements of Britain.

Discussing the long-running Ideal Home Exhibition's early days in the first half of the twentieth century, cultural historian Judy Giles argues that the exhibition

> has to be seen in the context of an increased emphasis on visual display and spectacle that is characteristic of modernity […]

Fig. 1.2 Ekco advertisement from the Festival of Britain catalogue (1951)

> Retail exhibitions were seen not only as venues to promote 'the retail sale to the general public of novel and popular commodities' but also as opportunities to display, often spectacularly, the products of science and technology.
>
> (2004: 110)

The spectacle of modernity was to be seen in almost every element of the Festival of Britain, and, in line with this, many of the BBC programmes broadcast from the Festival, produced by the Corporation's Outside Broadcast Unit, stressed this in their design. For example, the series *Festival*

Close-Up (1951) produced episodes such as 'Adventure in Architecture' (tx. 7/5/51), which focused on a series of slow pans up and across Skylon, the Radar Tower, Observation Tower, Lion and Unicorn and so on,[3] and 'Experiment in Design' (tx. 21/5/51), in which the designs of the Festival Pattern group were shown in great detail and lingering close-up. In the latter, the 'Haemoglobin' pattern produced by the Festival Pattern group was shown by cross-cutting between a silk scarf and an extreme close-up of a plate containing the presenter Patricia Laffan's meat ration as she explained 'Perhaps I better say that the colours are – well somewhat violently gay – atomically gay even' (BBC, 1951a).

The special programme *The Lights of London* (tx. 16/7/51) epitomised the Outside Broadcast Unit's presentation of the spectacle of modernity at the Festival. In the billing produced for the Programme Planning Department, Keith Rogers of the Unit, eloquently describes the programme:

> The Festival Lights bordering the Thames between Westminster and Blackfriars provide an unforgettable sight. Tonight, a visit is made to the South Bank at dusk to watch the scene as floodlit buildings assume a new beauty and the hardness of concrete and steel is softened or lost in the lace patterning of thousands of lights.
>
> (Rogers, 1951)

A similarly poetic description of *The Lights of London* was provided in a later press release:

> On Wednesday the television cameras are attempting to capture the magic of the scene as concrete and steel, and the stately utility of London's buildings, together with the inevitable untidiness and litter of a great exhibition fade with the twilight, to be replaced by pools of light rippling in the river, by the flashing fountains and the soft, almost ethereal, beauty of floodlit stone.[4]

The lyricism of this programme, with its languorous pans[5] over the South Bank at dusk, was underscored by a romantic orchestral soundtrack which included the Melanchrino Strings' 'Portrait of a Lady', Peter Yorke's 'Dawn Fantasy' and Easthope Martin's 'Evensong'.[6] During the period of the Festival then, outside broadcasts concentrated on emphasising the spectacular nature of the Festival and its associated architecture and imagery, alongside coverage of the pomp of the opening and closing ceremonies, and a special five-month season of broadcasting including 'a season of *Festival Theatre* [...] a series on the history of British entertainment in the twentieth century and a cookery series [which] featured regional British cooking' (Easen, 2003: 61).

In addition to these programmes that concentrated on the visual elements of the Festival of Britain, the two main television exhibits, the 'Television' pavilion and the adjoining 'Telekinema'[7] in the 'Downstream' section of the South Bank exhibition, were also designed to reflect on and demonstrate the spectacle of modernity. The former (Fig. 1.3) was a two-story building designed by Wells Coates intended to 'tell the whole story of broadcasting from the beginning' (Bishop, 1949). Malcolm Baker Smith, who wrote the treatment for the pavilion and supervised its content, was instructed by the BBC's Controller of Television, Norman Collins, to produce

> a broad, popular, historical treatment of Television from Baird days up to the present, with a peep into the foreseeable future, always remembering that everything in the script must be illustratable in terms of an object which can be placed on exhibition.
> (Collins, 1950)

The Television pavilion was fronted by an eye-catching 'shop window' containing a 14 foot by 8 foot 'penumbrascope',[8] on which was seen 'in constant motion and with a constant change of colours, the word "TELEVISION", together with stylised shapes representing the three main components of television – Light, Vision and Sound' (Baker Smith, 1950: 3).[9] Four large framed panels were added to the outside of the building just before opening with '"blow ups" of selected photographs of TV productions – "star" productions – to be permanently mounted in these frames, for the general view of the public' (Coates, 1951). The front of the building thus emphasised both the spectacle of television as 'show business' through the giant images of television stars, but also, through the inclusion of the penumbrascope, offered a visual representation of the technology of television as spectacular in and of itself.

The downstairs section, focusing on television's early history, displayed Baird's first working model, complete with its ventriloquist's dummy head;

Fig. 1.3 The Telekinema and the Television Pavilion at the Festival of Britain (1951)

photographs of the early TV pioneers at work; a model of Alexandra Palace with *Picture Page* (BBC, 1936–52) and a variety programme in production; a photomontage of television at Radiolympia; the story of the wartime close-down and restart told in words and pictures; and maps showing the expansion (both actual and proposed) of transmitter coverage. Upstairs, Baker Smith designed an exhibit to take the viewer 'backstage and explain the mysteries of the technique whereby entertainment is brought to him in the home' (Baker Smith, 1950: 8): this included a series of photographic panels representing the work of different departments at the BBC (the Outside Broadcast Unit, the Children's Department, Drama, and News) and various associated models, including a model of the *Andy Pandy* stage (BBC, 1950–2), plus, at the centre of this part of the pavilion, a mock-up studio through which the story of television production could be told. Whilst the production gallery was full-sized and populated by wire sculptures to represent the production team, the studio was a detailed, scaled-down model. Baker Smith had initially intended to stage an original television play called *The Traveller's Returns* in this studio, but Val Gielgud felt that the Drama Department was too busy to put time into its design and 'production', so instead John Dryden and Henry Purcell's opera *King Arthur, or the British Worthy* was used as it was being performed at the Albert Hall whilst the pavilion was being prepared in February 1951. Dryden's descendent, Sir Noel Dryden, provided the voice of the producer giving camera directions, which could be heard throughout this part of the pavilion. The opera, a more obviously 'spectacular' production than the originally planned teleplay, thus combined the cultural cachet of high art with a technological display of studio production techniques. The 'technical achievements' narrative that dominated the Television Pavilion was also echoed in the television section of the Transport and Communication Pavilion at the Festival, where various items of television technology –including cathode ray tubes, camera tubes, models of the BBC's proposed White City site and a 750ft aerial mast, plus various television sets, were displayed. However, whilst these exhibits placed television production and broadcast as the object of display, the Telekinema, which stood adjacent to the Television Pavilion, transformed television from a domestic broadcast medium into a spectacular big-screen show.

Also designed by Wells Coates, the Telekinema was a building in the 'modern' style, which was framed by the Abacus screen[10] that ran along the back of the building, tying the Telekinema into the futuristic 'atomic' design of the broader Festival (Fig. 1.3). Described by Becky E. Conekin as a 'memorial to the future' (2003: 55), the interior of the Telekinema was also innovative in its design: the projection room, which doubled as a television studio for short interviews, was placed behind glass and visible within the foyer. Barry Turner argues that 'For most visitors this alone was worth the price of admission, since it was the first time they had seen, as it were, behind the screen. Few, indeed, had ever seen a television set' (Turner, 2011: 210). The auditorium itself, which was staffed entirely by red-haired usherettes dressed in green uniforms designed by Hardy Amies, was divided into lower and mezzanine levels; it seated 410, with 252 in the stalls and 158 on the balcony. A perforated panel framed the screen, rather than the traditional curtained proscenium arch of contemporaneous film theatres (Fig. 1.4). This offered cleaner, sleeker lines in keeping with the

Fig. 1.4 Inside the Telekinema at the Festival of Britain (1951)

modern style, but the design also nodded towards the rectangular framing and harder lines of the television set, given that this space was to be used to project both film and television.

The inclusion of television in the Telekinema's programme is, in actual fact, contestable, given that from the outset, the 'television' elements of the programme were not, primarily, to be extracts of broadcast television or sequences filmed to be shown on broadcast television at a later or earlier date than their theatrical showing. Rather, they were live introductions to the 70-minute programme by the manager of the Telekinema, Frank Hazell, from the projection booth; these mimicked the presentational style of television with its direct address to camera and conversational tone, alongside the occasional short outside broadcast from some other part of the Festival, all managed by the British Film Institute rather than the BBC. These 'television' segments of the programme were then followed by sponsored documentaries and 'experimental' films which demonstrated stereoscopy (3D), colour and stereophonic sound.[11]

In the planning stages of the Telekinema, it is clear that J. D. Ralph, the BFI's film officer, had intended for the programme to include a greater proportion of broadcast television than was actually produced. As minutes from a meeting in March 1950 show, Ralph proposed that the BBC was to provide

> selections of its normal transmission of outside broadcasts on national sporting or other events. These programmes, broadcast from Alexandra Palace, would be picked up by aerial on the cinema or on the Shot Tower. Regular BBC outside broadcasts originating on the South Bank could [also] be transmitted to the cinema if convenient by cable.
>
> (Ralph, 1950a: 1–2)

However, the BBC was reluctant to participate on three counts. Firstly, there were distinct worries that participating in the Telekinema would somehow degrade the BBC's reputation as a public service broadcaster, through an association with the more commercial/populist aspects of the British film industry. As early as December 1947, the head of the Television Service, Maurice Gorham, hand-wrote a note at the bottom of a memo from Kenneth Adam (Director of Publicity) about the Telekinema stating: 'You are right. It is most important that Television should be kept entirely distinct', in response to Adam's suggestion that 'we would not want to appear as an appendage of Rank and Korda' (Adam, 1947). Whilst these fears turned out to be unfounded (a different version of British cinema than anticipated here was ultimately offered to visitors to the Telekinema through its emphasis on documentary), it is clear that the BBC wished to distinguish itself from the Ranks and Kordas of the British film world.

The second worry, and perhaps a more valid one, was that showing BBC programmes as broadcast would make television look like the inferior cousin of cinema, given that its programmes were designed for a much smaller screen. The minutes of a 1949 planning meeting record that

> Mr [Norman] Collins [Controller of Television] stressed that it had also to be borne in mind that BBC television productions generally were devised for reproduction on the small screens of domestic receivers and would not be aesthetically satisfactory for larger screen reproduction. On the technical side the large-screen reproduction of 405 line television would lead to invidious comparisons between television picture quality and cinema picture quality.
>
> (BBC, 1949a)

Whilst the programmes the BBC planned around the Festival were frequently spectacular in nature, images produced for the small screen would still look decidedly less spectacular and less crisp and sharp when viewed alongside colour 3D images or the crisp clarity of the cinematography of Humphrey Jennings and Basil Wright et al. However, again, these fears appear to have been somewhat unfounded. For the just short of half a million visitors to the Telekinema during the Festival,[12] the opportunity to see 'television', many for the first time, superseded any worries about disparity in picture quality. According to Barry Turner (2011: 214) the biggest hits of the Telekinema were the demonstration of 3D and the opportunity to 'see yourself on television'. Frank Hazell, General Manager of the Telekinema, wrote the following to Cecil McGivern, Controller of Television at the BBC, towards the end of the Festival, also reflecting on the success of television's appearance in the Telekinema:

> As you are no doubt aware, the Telekinema has proved one of the outstanding attractions of the South Bank, and to date nearly half a million people have seen performances in this theatre. As the most impressive thing to them is the large screen Television we feel that at the final performance we should try to do a fairly large production, and this letter is to ask your blessing and help in obtaining the services of some of the better known BBC personalities [...] I feel it would be a grand winding up programme for what must have been a big selling point for British television.
>
> (Hazell, 1951)

However, the final, and almost insurmountable, problem for the BBC in showing their programmes in the Telekinema were the issues raised by the Association of Cinematograph Television and Allied Technicians and Equity. J. D. Ralph had initially underestimated the complicated problem of negotiating with these powerful unions, which objected to BBC programmes being shown to a *paying* audience,[13] as expressed in the 1949 planning meeting minutes: '[Norman] Collins pointed out that unresolved copyright and union difficulties at present entirely prohibited the large-screen showing of BBC television programmes involving Professional artists, certainly if a charge was made for admission to the tele-cinema and possibly in any case' (BBC, 1949a). Following this meeting, Ralph began a protracted correspondence with W. L. Streeton (the BBC's Head of Contracts) in order to try and find a way around the problem. Streeton eventually managed to clear several programmes, including some newsreel about the Festival, some football coverage (which had to be cleared with the Football Association) and 'that portion of the *London Town* programme to be transmitted in our Television Service on 3 August which concerns the Telecinema itself and which involves certain staff of the Telecinema and Mr Richard Dimbleby' (Streeton, 1951). The term 'rediffusion' was used to describe this process, which to a certain extent mirrored earlier experiments with transmitting television into large screens in theatres and cinemas.[14]

Reporting back to Sir Henry French of the British Film Producers Association about television at the Telekinema, J. D. Ralph summed up the BFI's position:

> I would say that our policy on television in the cinema is being governed by the following principal considerations: (a) That cinema television is already a practical proposition from the technical point of view and has been demonstrated in various forms in this country and the USA. British engineers have made a substantial contribution in this field and in view of the Festival's desire to present not only present-day achievements but to foreshadow the future, a serious effort should be made to include this new medium of entertainment in the Exhibition. (b) That the Festival has a responsibility that goes beyond the terms of reference of the BBC in the matter of television. I refer

to our responsibility for demonstrating, if possible, the type of equipment which our industry is building for the home market and for export overseas. (c) That it is essential to maintain a close working relationship with the BBC on all matters relating to the television programming for the Telecinema. Our main objective, however, is not the rediffusion of the BBC's regular programmes which are designed primarily for the home viewer.

(Ralph, 1950b)

So what was it that was being shown as 'television'? Was it really television? Should it be understood as such in histories of the medium, despite being narrowcast on a closed circuit rather than broadcast via a transmitter? Although the BFI's production of this kind of programming at the Festival of Britain was a prominent example, it was in fact common practice for this site-specific programming to be produced in and for large-scale public exhibitions. This 'public television' might therefore be best understood as a third category, separate from home broadcast and cinema, with different kinds of content and styles which in many ways might be more closely associated with vaudeville and variety performance, and which might be easily aligned with the spectacular.

By far the most popular aspect of the Telekinema's programme was the opportunity to 'see yourself on television', where the cameras were turned on the audience and people could thus see themselves, or their friends and family, on the big screen. In actual fact, this exhibition technique had been used since the early 1930s as a way to demonstrate the unique properties of television. For example, in the 1930 *Ideal Home Exhibition* catalogue, the copy for the Baird Company's stand at the exhibition reports on the possibility for visitors to Earls Court to be televised:

> The Baird Commercial 'Televisor' is on show for the first time in the history of electrical science [...] the transmitter may be seen at 195 Hammersmith Road where a further display of 'Televisors' and component parts are on view. Any person wishing to be 'televised' should apply to the Stand for a ticket which will entitle them to go to the transmitting end.
>
> (1930: 304)

Whilst it was impossible to 'see yourself' at the 1930 Exhibition, this highly popular exhibit, in broadcasting an image of the participant back to companions in Earls Court, emphasised television's space-binding properties, its potential as a medium of direct address and its startling sense of a spectacular form of intimacy. This exhibition technique had also been popular in the US in the pre-war period. As William L. Bird Jr describes, in the General Electric, Westinghouse and RCA exhibits at the 1939 World's Fair 'fairgoers could see themselves on closed-circuit television. Some visitors recorded their friends' television appearance in snap shots, and exhibitors issued souvenir cards to their guests, certifying that "– has been TELEVISED"' (1999: 186).

Time and again, those staffing television exhibits with a 'see yourself' element reported back that the opportunity to see oneself on screen was endlessly fascinating for visitors to the BBC's stands. To give an example, at the Scottish Radio and Television Exhibition of May and June 1957, held at Kelvin Hall in Glasgow, the BBC Publicity Officer for Scotland, D. E. Stewart, reported that 'the thing that really mattered, above all else at the exhibition, was the "See Yourself, Hear Yourself" equipment' (Stewart, 1957). He elaborated:

> Crowds will look at pictures, three dimensional models, or demonstration machines in passing, they will pause at a sales counter, and in special cases talk for a few minutes at an inquiry counter. But they will be held for half an hour by a panning 'see yourself' television camera working on closed circuit and a 'hear yourself' tape recording machine worked in conjunction with a roving microphone and a local PA system on the stand.

Stewart went on to recommend that any future BBC exhibits at fairs, exhibitions and expos should be designed around such an exhibit. Such an exhibit, holding the viewer's attention for a much greater length of time than anything the designers could construct, offered the opportunity for the visitor to literally 'make a spectacle of themselves', to put themselves on screen. Like those who mug and grin behind a news reporter delivering a piece to camera on location, the 'see yourself' exhibit extended the pleasure of being transformed into a televised image, the pleasure of *becoming spectacle*.

At the Festival of Britain, the BBC staged its own exhibition that drew on the 'see yourself on television' conceit. The BBC exhibit, at their studios housed in an Edwardian commercial property at 201 Piccadilly, combined models and pictures of BBC's radio and television production with a closed-circuit television setup that Head of Publicity, Douglas Ritchie, had to fight hard for, and which meant that an admission charge had to be paid for this exhibition in order to cover the associated costs. Leonard Potter (the BBC's Exhibition Designer) reported that 'In the first six days of the exhibition, where the public can record and hear their own voices and see themselves on a television screen, there have been some six thousand visitors' (1951); later in the Festival's run, Ritchie reported average visitor numbers of 1,600 per day, rising to over 2,000 in a week in the middle of the exhibition (1951). Clearly, one of the main attractions of the exhibit was the opportunity for visitors to 'see yourself on television', and it is certain that people's interest in this was further piqued when, at 8pm on Saturday 18 August, the BBC broadcast a 'surprise item' from the Piccadilly exhibition, with Sylvia Peters commentating on a live feed of visitors to the exhibition, encouraging them to send messages home or to 'wave for the viewers in Birmingham' and so on. Following the broadcast, people began to come to the exhibition and do 'turns' in front of the camera; just two days later, the BBC sent out a press release entitled 'Beginners Please' which described the following:

> Into the BBC's Piccadilly Exhibition today walked a red haired youth with a paper carrying bag. It was the fairly slack hour between tea and dinner and as it happened there was nobody standing in front of the camera which televises visitors and at the same time shows them how they look on screen. The youth drew from his bag a pack of cards and other impediments and began to practise various tricks, for he was an amateur entertainer who wanted a television audition and thought that to practise before the camera and see how he would appear to viewers would be a good thing. Watched by interested visitors, he went through his little set, turned this way and that to give people the best of his rather snub-nosed profile and then, with a polite 'Ta' to the cameraman, repacked his paper bag and left the exhibition.
>
> (BBC, 1951b)

Here we clearly see the roots of what would become the TV talent show five years later – 'ordinary' people 'making a spectacle of themselves' on television – though as Su Holmes points out, the 'talent show has a complex, cross-media history, given that amateur talent competitions were part of music hall and other leisure contexts such as pubs, holiday camps, "end of pier" contests, and working-class cinemas in the 1930s before the advent of broadcasting' (2014: 25). More broadly though, the 'see yourself on television' exhibit continued to be a feature of the BBC's public exhibitions around the UK: on the BBC stand at the National Radio Show in 1954, for example, visitors could buy a filmed copy of themselves on screen, and this was still a popular feature on the BBC stand at the 1962 Ideal Home Exhibition and the ITV stand at the same exhibition in 1965.

The spectacle of television production at the National Radio Show

The other key exhibition in the UK where television's technological achievements were lauded alongside the presentation of the medium as a site of spectacular, populist entertainment was Radiolympia, or the National Radio Show. Whilst, as the title suggests, the exhibition started out as a trade and industry exhibition for radio, it soon incorporated television, particularly following the restart of television in 1946. At this point the Director General of the BBC, Sir William Haley, was keen that television did not overshadow radio at Radiolympia 1947,[15] despite the fact that the set manufacturers wanted television to be a key focus of the exhibition. In the end, 24 television manufacturers confirmed for the exhibition, 39 sets were on display and a 250 foot-long exhibit allowed the public to wander along watching these sets.

But what was being broadcast on these sets? Files at the BBC reveal that sets in the exhibition showed a combination of demonstration films provided by the manufacturers' associations, broadcasts from Alexandra Palace and, in the most prominent place in the test schedules, programming which was made live on stage at Radiolympia and simultaneously narrowcast to the sets dotted around the exhibition and broadcast via an outside broadcast link up to the viewers at home. These programmes included *Café Continental* (BBC, 1947–55), *Geraldo and Orchestra* (BBC, 1946–7), *Picture Page* (BBC, 1936–52), *Cookery with Phillip Harben* (BBC, 1946–51), *Stars in your Eyes* (a variety programme , BBC, 1946–50), *The All-Coloured*

Variety (BBC, 1947) and *Come and be Televised from Radiolympia* (BBC, 1947). A BBC press release describes this programming:

> As the newest form of broadcast entertainment, television takes precedence, and the whole of the large stage is designed primarily as a television studio, from which sound and vision will be radiated daily to the tens of thousands of viewers in London and the Home Counties [...] you will be looking at a studio which is a big-scale version of the studios at Alexandra Palace [...] Outstanding among shows specially devised for television will be a large-scale production of that acknowledged favourite, *Café Continental* [which] [...] will be a lifelike revival of the exotic French café show which entertained the troops in Cairo. Another big production which will absorb all three stages will be an *All-Coloured Variety* programme bringing together some of the best coloured artists now performing in this country.
>
> (BBC, 1947)

It is clear to see here that a great emphasis was placed on broadcasting and narrowcasting big, exotic, spectacular shows when television met the exhibition immediately postwar and beyond. Given that for many people, Radiolympia was to be their first encounter with television, the impression given of the medium was of spectacular glamour, ritzy show business and an emphasis on the large scale, rather than the small. This, of course, runs counter to the delineation of television as a medium of intimacy in the same period (see Jacobs, 2000), though it is significant that whilst some genres thrived in this setting (variety, dance and music shows, even some demonstrations and interviews), drama (the subject of Jacobs' study) was not part of the Radiolympia roster. Television at the exhibition thus most closely resembled theatre and vaudeville, and time and again in-house productions emphasised the demonstration of scale and technical achievement in the production process.

It is important to note that these productions catered to multiple audiences watching in very different ways, hence the emphasis on variety and vaudeville: the audience in the 450-seat auditorium in front of the stage, the potential buyers perusing the television sets in the exhibition – who needed to be impressed enough with sound and picture quality and what could be done on television to invest in it as a new home entertainment technology – and the viewers at home. This raises interesting issues about address: whilst the early adopter would have been almost certainly upper or upper middle class (given the expense of sets at this time), variety was more commonly associated with lower-middle-class tastes at this time. Thus the BBC worked hard to suggest a sense of sophistication and exoticism in their variety programming (think *Café Continental* rather than *Sunday Night at the London Palladium* (ATV, 1955–67; 1973–4)): this was, initially at least, the entertainment of the colonial club rather than the music hall. As the postwar Radiolympia exhibitions continued, productions got more and more elaborate and technically challenging to produce. In 1949, for example, the BBC produced both the *Radiolympia Follies* and the *Icecapades*, an ice show with a large programme budget of £220, which was justified by the fact that it was felt that it provided a 'spectacular opening' to the Radiolympia coverage (BBC, 1949b). The minutes of the planning meeting of the 1949 exhibition explicitly state that the aesthetic of these programmes primarily served the studio audience: 'television shows planned are spectacular rather than intimate in type, in order to appeal to the viewers in the gallery as well as the actual televiewer' (ibid.). However, these programmes were also viewable from a fourth space, a glass-panelled gallery which flanked the studio, as well being watched within the auditorium and on television sets inside and outside the exhibition, as described in the 1949 catalogue:

> From a glass-panelled gallery skirting the studio at a height of 25 feet you can look down on a brilliantly lighted stage, covering 7,200 square feet, on which spectacular presentations such as *Café Continental, Music Hall, Ice Parade* [sic] and *Grand Ballet* are being radiated at the moment of performance to more than 150,000 television receivers in London and the Home Counties.
> (BBC, 1949c)

This offered a doubling of spectacle in which the process of production becomes spectacular in itself, alongside the spectacle of the presented variety shows. The audience viewing in the gallery had a prime view of the 'workings' of television production (lighting rigs, cameras, sound recording equipment and the production team working 'behind the scenes'), and were thus encouraged to appreciate the spectacle of scale and technical achievement.[16] Later, in 1952, the BBC announced that they had built 'the largest sound and television studio ever made' (BBC, 1952) at the National Radio Show, big enough for the large-scale spectacle of a 'Corps de Ballet of 16 dancers and a chorus of 16 singers, in a Viennese setting' (ibid.). The following year, this space was adapted again to fit a cinema screen above the proscenium 'so that the audience of nearly 1,000 may see not only all that happens on the studio floor, but also the picture chosen for transmission over the air or around the closed circuit within the exhibition' (Radio Industry Council, 1953). This image, of a 'blown up' television screen of cinema-style proportions hanging above the proscenium arch of a theatrical stage used as the set of a series of television programmes encapsulates the idea that television in the mid-twentieth century was a hybrid medium, particularly in its presentation of spectacular entertainment. Here we see an entertainment form which is simultaneously 'television', 'film' and 'theatre'.

In later years, however, the spectacle of television at the exhibition shifted away from putting television engineering 'on show' and more towards the showcasing of the television celebrity and a focus on popular programming. This was almost certainly prompted by the advent of commercial television in the UK and by the promotional strategies of the BBC's competitors, the ITV companies. At the Scottish Radio and Television Exhibition of 1957, Deputy Director of Television Broadcasting, Cecil McGivern, reported that

> our stand, in my opinion, lacked the necessary flamboyancy and touch of showmanship for exhibitions like this [...] The Scottish Television Ltd. stand was used more as a concert party platform [...] Not only were personalities interviewed but performers performed, singers sang and comedians 'gagged' to the audience standing round. This is crude but nevertheless very cheap and quite [an] effective substitute for studio or arena.
> (McGivern, 1957)

It was thus no surprise that months later the BBC produced a celebrity dais as the centre of their exhibit at the National Radio Show, described by its designer, Austin Frazer, as being

> intended to convey the impression of 'glittering spectacle' – like the first sight of a birthday cake, highly decorated and scintillating with radiant candles. The content will be exciting, it will have the maximum popular appeal, and the main emphasis will be on personalities and audience participation.
>
> (Frazer, 1957)

We therefore see a shift in the tone and style of the television exhibit in the latter years of the 1950s, away from an emphasis on technological achievement and cosmopolitan variety programming, towards a focus on the spectacle of celebrity. These exhibitions produced the particular frisson of proximity to celebrity in a way that fulfilled the (previously unfulfilled) promise of television to bring you 'close up' to star performers. If television at home offered an imagined intimacy with the stars, television at the exhibition offered a more tangible, more real encounter with stardom.

In the following year, this sense of proximity to the excitement of 'show business' was realised by the ITV companies through an immersive exhibit that gave the exhibition visitor a fully embodied experience of the excitement of television:

> [The Independent Television Wonderland], staged by Associated Rediffusion and Associated Television in conjunction with the Independent Television Authority, Independent Television News and the *TV Times,* has been designed to transport spectators into a realm of fantasy, colour and excitement. Through a screen surrounding the entire stand, on which are depicted scenes from Lewis Carroll's beloved masterpiece, the visitor enters a dimly lit vestibule and passes on into another foyer, wrapped in complete darkness, which leads into the various halls. There are seven halls, each devoted to a particular aspect of independent television and all brightly illuminated – the idea being to heighten the sense of wonderment and to make the spectator feel that he has stumbled from a world of unreality into settings made familiar by a year's viewing.
>
> (Anon., 1958)

These halls included the Quiz Hall,[17] the Display Hall,[18] the Hall of Music,[19] the Hall of Drama,[20] the Wild West Hall[21] and a Hall of Personalities, in which the visitor 'can see not only those who already provide his entertainment during the year but also, through various talent contests, those who might do so in the future' (ibid.), as well as being able to 'see themselves on television' via another closed-circuit TV exhibit. This immersive exhibit, which presented television as a dizzying wonderland of entertainment and celebrity that visitors could enter, clearly emphasised the medium's spectacular qualities in order to attract viewers to the nascent commercial television service. It also paved the way for other immersive spectacles of technological wizardry at the Ideal Home Exhibition.

The spectacle of domestic modernity at the Ideal Home Exhibition

Exhibits at the Ideal Home Exhibition also presented television as a spectacular medium. As I have argued (Wheatley, 2016), the female consumer-citizen was targeted by broadcasters and set manufacturers alike through the complex and varied TV exhibits of the mid-century exhibition, particularly at the Ideal Home Exhibition. The spectacular nature of television's role at the Exhibition thus needs to be understood in relation to theories of domestic modernity, and the ways that modernity was lived, expressed and imagined in the private worlds of women. Previously, I have traced a history of television as a gendered object at this exhibition, analysing the place of television as it was presented in both 'ideal' and 'future' homes, and tracing a history of an address to women in broadcasters' and set manufacturers' exhibits. The ideal home was, of course, a central concern for the postwar government in the UK. In discussing the presentation of the home at the Festival of Britain, for example, Harriet Atkinson reflects on the ways in which exhibits at the Festival echoed the recommendations of the 1944 Dudley Report on the *Design of Dwellings* and the subsequent *Housing Manual* and their focus on producing 'ideal family environments, consulting women as housewives and mothers on how these new homes should be designed, concluding that improved conditions would be important in promoting "family life" and encouraging families to grow' (Atkinson, 2012: 160). However, in addition to explicitly addressing women and their role in the postwar reconstruction of family life, exhibits at the mid-century Ideal Home Exhibition also focused on what we might understand as the 'spectacle of modernity'. This chapter concludes, then, with a consideration of the exhibitors' role in constructing this spectacle of modernity at Earls Court.

In her pictorial history of the Ideal Home Exhibition, Deborah S. Ryan describes the way in which the Exhibition sought to educate the general public 'in the latest labour-saving ways and [entertain] them with fantastic spectacles which preserved home-making as both scientific and glamorous' (1997: 12), arguing that the Exhibition needs to be seen as a constituent part of a broader 'culture of the spectacular' (ibid. 13) in the early and mid-twentieth century. The Exhibition was first established by the *Daily Mail* newspaper in 1908 and continues to be staged annually, now back in the Olympia exhibition space in London.[22] In the mid-twentieth century, its visitor numbers rose to a high of 1,329,644 in 1957.[23] From its earliest years, its goal has been to bring together everything associated with having an 'ideal home': the most up-to-date inventions for the modern house, and the showcasing of the latest housing design. Looking at this exhibition, then, tells us as much, if not more, about the history of twentieth-century Britain and the tastes and aspirations of ordinary families, as looking at the much more 'chic' Festival of Britain. As Ryan argues:

> The Ideal Home Exhibition has told a different history of the domestic interior to that of the tasteful galleries of decorative arts museums. It has presented a design history that largely rejected the Arts and Crafts movement and Modernism, but still embraced modernity [...] Despite, or perhaps because of, its undoubted appeal, cultural commentators have often dismissed the Ideal Home Exhibition. Many have felt uncomfortable with its reflection of consumer aspirations for goods and lifestyles that they have thought of as frivolous and empty [...] Yet the history of the Ideal Home Exhibition is the history of the hopes, dreams and aspirations of the respectable working classes and middle classes, or conservative and ordinary people.
> (ibid.: 19)

Television was at the centre of this narrative of consumer aspiration and affordable luxury at the Ideal Home Exhibition from the very earliest days of the medium.

In the prewar era of the Exhibition, much focus had been simultaneously placed on glass and electricity as the twin signifiers of domestic modernity. This is particularly encapsulated in the 1930 'Pavilion of Light', an all-electric 'house of the future which was constructed almost entirely out of glass, complete with walls that shimmered with coloured light. This 'thrillingly modern technology' (ibid.: 62) was also in evidence in 1938 when a 'City of Glass' and an 'All-Electric House' were advertised as highlights of the Exhibition. The potential of the glass structure to provide both 'stunning' views *from* the home, as well as spectacular views *of* the home, and the promise of electricity in bringing modernity *to* the home, looks forwards to what television was shown to offer at the Ideal Home Exhibition. As a material object made largely of glass and electrical components, the television set might be seen to manifest or encapsulate the excitement of this modern consumerism in the 1930s, as well as the potential space-binding properties of both. This argument is similar to Lynn Spigel's discussion of television's place in suburban home design of the 1950s, in which she proposes that 'the domestic architecture of [this] period mediated the twin goals of separation from and integration into the outside world' (1997: 212); in this work, Spigel focuses on the role of glass, and the picture window in particular, as spatial/architectural metaphor for television's potential to draw together public and private.

Whilst television's space-binding properties were initially highlighted at the Exhibition by Baird's 'Televisor' exhibit in 1930, discussed above, rendering the medium as pseudo-telecommunications device in its earliest demonstrations, in the first full year of broadcasting it was television's ability to bring exciting news and sporting spectacles *into the home* which was emphasised at the Ideal Home Exhibition. In the foreword to the 1938 Exhibition catalogue, television is described as a 'miracle the future effects of which none can adequately foresee' and visitors are invited to 'exercise your judgement upon this great achievement and to speculate about the future in your own home when the world's events are whirled before you at your fireside in colour' (WWH, 1938: 9). Later, in the same catalogue, a Marconiphone Television advertisement states:

> The Ideal Home to-day has Television because to be really up-to-date and to enjoy all that is going in this very wonderful

> modern world you *must* have Television. There is no more thrilling entertainment than having your own stage and your own cinema screen in your own home. Think of exciting events you are missing – Wimbledon tennis, Championship fights, the Boat Race, the Lord Mayor's Show etc. etc. – seeing and hearing them all just as they are happening.
>
> (1938: 53)

Here television's identity as a device of modernity is tied to its ability to bring pageantry and sporting spectacle into the home, all in the context of an encounter with the future, or certainly the modern, via television. Citing Mica Nava's work on the department store as a key site of modernity in the early twentieth century, Judy Giles has argued that it was 'in the exploding culture of consumption and spectacle, symbolised by the department store and the exhibition, that the daily lives of ordinary women were most touched by the processes of modernity' (2004: 104–5). At the Ideal Home Exhibition, an encounter with television was frequently figured as an encounter with the spectacle of modernity.

Whilst it might seem unsurprising that this presentation of television as a device of futuristic modernity was offered in the 1930s when television was still very much a novelty, a spectacular object in and of itself, it is perhaps more surprising to hear that this presentation of television persisted into the 1960s at the Ideal Home Exhibition. For example, in 1961, the BBC's stand at the Ideal Home Exhibition presented television as an exciting, futuristic technology by guiding visitors through an immersive sphere in which they would encounter 'television', presumably taking their lead from ITV's Independent Television Wonderland at the National Radio Show in 1958. It is a shame that no images of the 1961 BBC stand have survived; however, evocative descriptions in the BBC Publicity Department files paint a vivid picture of this exhibit:

> 'An Aladdin's cave of electronic delights' might be the description of the BBC stand. Within a spherical structure, 36 feet in diameter, 12 feet in height, a cascade of colour, light and sound tells the story of the BBC. It is a world of vision and sound bringing to visitors to the Exhibition the history of BBC Television in its twenty-fifth anniversary year.
>
> (Thomas and Campney, 1960)

A later press release expanded on this description:

> Visitors enter a darkened arena and will then take a three minute journey through the many aspects of BBC broadcasting, described in graphic illustration brilliantly lit. We have attempted, without using the written word, to provide the visitor with a narrative about the BBC. There will be a taped commentary [by Derek Hart] and great use will be made of ultra-violet light.
>
> (Campney, 1961)

Further, a memo from Perry Guinness, the BBC's Publicity Projects Organiser, to heads of programmes that might cover the Exhibition explains more about the content of the immersive sphere:

> Points during the talk will be given sudden and dramatic emphasis by sound and lighting effects and by moving models appearing from the darkness. [Hart's] talk will end with reference to the weather. A full scale thunderstorm will be developing within the sphere as the public moves toward the exit.
>
> (Guinness, 1961)

The models in the sphere were symbolic images, rendered almost phantasmagoric via ultraviolet light, which represented different areas of BBC programming: 'Gardening programmes for instance are represented by a flower suddenly bursting into bloom, schools programmes by a huge alphabet floating in space' (Patterson, 1961). The sphere was subsequently described in the Radio Newsreel as 'like a ghost train [...] vivid, alive and gay' (Radio Newsreel, 1961). These descriptions thus tell the story of an immersive exhibit in which television (or the story of television, at least) was experienced as *kinaesthetic* spectacle. Just as 'meet the stars' exhibits at the National Radio Show had offered the opportunity for viewer-visitors to fulfil television's aesthetic promise of intimacy with the television performer, so this 'dome of wonder', and the BBC's travellator exhibit the following year that provided 'a kind of magic carpet from which visitors can see many of the faces of people who have made BBC programmes memorable',[24] attempted to symbolically fulfil television's promise of unfettered mobility. Again, historical work on US television has identified similar narratives within promotional strategies for television: William Boddy's discussion of DuMont's print ads promising their viewers the opportunity to 'be an armchair Colombus on ten-thousand and one thrilling voyages of discovery!' also highlights this rhetoric of 'mobility, omnipotence, and adventure' (2004: 54), though these immersive exhibits at the Ideal Home Show attempted to transfer this promise into experience. Here, television is presented as a technological marvel, a vision of light and colour; it is taken out of its quotidian, domestic context and translated into an immersive spectacle. As far into television's history as the early 1960s then, these exhibitions positioned television as the epitome of spectacular modernity.

The legacy of television at the exhibition

So what is the legacy of television at the mid-century exhibition and where does this history take us in understanding television as a medium of spectacle? Evident in this history is that those involved in promoting television in its first three decades focused on its spectacular nature; their promotional strategies at the mid-century exhibition stressed the visual pleasures of live television performance, whether that be through exhibits which expressed the impressive scale and 'technological wizardry' of television production, the opportunity to 'make a spectacle of oneself' onscreen, or which foregrounded the presentation of large-scale variety acts, simultaneously performing for broadcast, closed-circuit and theatrical audiences. In the UK and US, this latter category of programming has largely receded into broadcasting history, save for the highly popular 'Got Talent' franchise produced by Syco Television for ITV1 as *Britain's Got Talent* (2007–) in the UK and *America's Got Talent* (2006–) for NBC in the US.[25] These programmes revisit the variety show format, featuring a series of singers, dance troupes, magicians and dancing dog acts who perform on a proscenium stage for a set of judges and a diegetic audience sitting in a theatrical auditorium; like the spectacular shows staged at the mid-century exhibition then, the acts must appeal simultaneously to a number of audiences – here the domestic viewer, the diegetic audience in the theatre and the judges watching in the auditorium. *Britain's Got Talent* thus plays a significant part in the revival of the TV Spectacular, weekend event programming in the UK known in the industry as the 'shiny floor show'.

Perhaps the other key legacy of the mid-century exhibition has been the development of events and screenings which in various ways position television as a spectacular form of public entertainment, alongside its status as a more intimate, domestic medium. I refer here first to the public, mega-screen exhibition of television, which reached its peak in the UK with the broadcast of the 2012 Olympic Games coverage on 'big screens' in cities across the nation, as discussed in the Conclusion of this book. For example, Park Live, sponsored by British Airways, was the flagship London 2012 'big-screen' site within the Olympic Village in Stratford; it formed part of a programme of nearly 70 live sites across the United Kingdom during the London 2012 Olympic Games, and hosted over a million viewers. Other outdoor sites included Victoria Square in Birmingham, the Millennium Squares in Leeds and Bristol, Festival Square in Edinburgh and Millennium Place in Coventry, alongside big screens in shopping centres like the Chapelfield Shopping Centre in Norwich. These intermedial screens, neither 'television' nor 'cinema' nor site-specific, closed-circuit 'footage' but rather a combination of separate elements of all three, offered a public, collective television viewing experience which in some ways replicated the big-screen viewing of the National Radio Show. Here the unusual scale of the television broadcast as well as the opportunity to view broadcast television collectively were the main attractions of these screens.

Second, the theatrical or site-specific stage show of a pre-existing television programme also forms part of the spectacular geography of public television, a legacy of the 'shows within the shows' that were produced as part of the mid-century exhibition. These range from specially produced, site-specific programming (e.g. *Blackadder Back and Forth* at the Millennium Dome, 2000), arena-based spectaculars inspired by, but not featuring, television programming (e.g. *Walking With Beasts Live*, 2007–; *X-Factor Tour*, 2005–; *Strictly Come Dancing Live*, 2007), theatre-based shows featuring actors and presenters familiar from television, often alongside live music and dance (e.g. *Horrible Histories Prom*, 2011; *Milkshake Live*, 2008–; *Little Britain Live*, 2005–7) and BBC Prom events which incorporate big-screen presentation of episode extracts, live performance by actors and presenters associated with each programme, and live orchestral/choral performance (*The Blue Planet Prom in the Park*, 2002; *Dr Who Prom*, 2008/2010/2013). What is interesting about these shows is that they test the limits of our understanding of the medium; they are made of television, but are not television (even though many of these shows are in fact remediated into 'special' broadcast television programmes, particularly the prom events[26]). However, they all suggest, like the television exhibits discussed in this chapter, that television lends itself to spectacular theatrical presentation, that its performers, its humour, its special effects and visual imagery are translatable to a theatrical stage, and that sometimes television programming itself can withstand being presented on hugely expanded screens. Whilst these stage shows derive (mainly) from television's more spectacular genres (e.g. natural history programming, children's television, science-fiction series, 'shiny floor' shows), and not all television programming would translate to the theatre or the arena, this says something about the spectacular aesthetics of a great deal of contemporary television.

And what of television at the exhibition? Is the medium still positioned at the centre of all that is new and modern? Do we still find television at the heart of the 'ideal home'? In March 2015, the Ideal Home Show returned to Earls Court, still aiming to present all that is new and exciting in the world of home and garden design. An analysis of this exhibition reveals a number of interesting things about the current status of television. On the one hand, it was clear at this exhibition that television stills plays a part in defining our domestic life: across the branding of different sections of the exhibition, well-known television personalities represented different elements of the exhibition's focus (interiors, gardens, home building, food and drink, etc.) (see Figs 1.5 and 1.6), tapping into the emphasis on domestic lifestyles *on TV* and acting as shorthand for the aspirational consumerism which television scholars have discussed in relation to TV since the late 1990s (e.g. Moseley 2000; Brunsdon et al. 2001). The enormous images of Gregg Wallace, Laurence Llewellyn Bowen, George Clarke and Suzi Perry around Earls Court rendered the television personality spectacularly present, as did demonstration daises within the space around which crowds gathered to view the TV performer plying their trade in the flesh. These demonstration areas have changed little since Fanny and Johnnie Cradock appeared on stage with their 'Kitchen Magic' show at the 1957 Ideal Home Exhibition, combining on-stage performance and an address to a live audience with large screens relaying close-up detail from this performance on the side.

Television comes to town

Fig. 1.5 The television personality as lifestyle branding at the Ideal Home Show, 2015: Laurence Llewelyn Bowen

On the other hand, an examination of the ideal homes at the centre of the exhibition revealed the absolute absence of television. In the three show homes at the centre of the exhibition inside Earls Court, there wasn't a single television on display. Here living room furniture had been organised around a feature fireplace or key lighting, and media technologies when present signalled a moment before the advent of television: the only broadcast technology on display in any of the three show homes was a 1950s-style retro Bush radio in the kitchen of the 'Gap House'. This suggests a nostalgic retreat from modernity, or certainly from the primacy of the home screen as the locus of modernity throughout the twentieth century. Perhaps the inference here was also that our screens have become personalised, that we have moved to the iPad rather than the shared family viewing of the domestic set, and that rather than being a spectacle of modernity in itself, the television set has receded into the background, hidden out of sight (as with the TV mirrors, an add-on technology to hide the wall-hanging set behind a 'mirror', on sale at the Ideal Home Show and discussed at greater length in Chapter 5).

Fig. 1.6 The television personality as lifestyle branding at the Ideal Home Show, 2015: Gregg Wallace

Outside of these ideal homes, however, the only media company present at the exhibition, Virgin Media, did offer a mocked-up living space on their stand in which television played a prominent part, located at the centre of the living room and on the kitchen wall, converging with computer technologies to offer the promise of limitless programme choice, superfast broadband and unprecedented access to social media whilst you view (see Fig. 1.7). The TV screens in this exhibit were then very present and promoted as offering never-before-seen clarity on which to enjoy the spectacles of contemporary television; television was not just television in this Ideal Smart Home, but part of a fantasy of totally immersive entertainment which warranted bigger, clearer screens with crisper, sharper, louder sound. Thus the 2015 Ideal Home Show demonstrates both continuity with and divergence from the history of the presentation of television at the mid-twentieth century. The simultaneous absence of television from the ideal home in 2015, and its presence as a domestic medium of 'never-before-seen' clarity of sound and image, speak of

Fig. 1.7 The Virgin Media 'Telly Heaven' Smart Home at the Ideal Home Show, 2015

the current uncertainty of television's future. When television is present at the exhibition, it is undoubtedly associated with the medium's more spectacular forms.

2

Spectacular colour? Reconsidering the launch of colour television in Britain

Chromophobia and the problem of colour

In 2000, the artist David Batchelor published a persuasive book on what he understood as the *problem* of colour, or 'Chromophobia', in which he argued that

> In the West, since Antiquity, colour has been systematically marginalized, reviled, diminished and degraded […] Colour is made out to be the property of some 'foreign' body […] [and is] relegated to the realm of the superficial, the supplementary, the inessential or the cosmetic.
>
> (2000: 22–3)

This book has become significant in the recent flurry of work on colour in film, not least to explain the relative lack of attention that has been paid to colour within film scholarship. Whilst there are a number of other attendant arguments which explain the dearth of work on film colour (about the subjective nature of colour perception, the potential unevenness of reproduced filmic colour in different settings and media and so on), Batchelor's work points us towards what might be seen as an underlying snobbery against a consideration of the aesthetic qualities of colour in cinema. Certainly this is how it has been interpreted within film studies.

Spectacular colour?

Chromophobia, or a deep suspicion about the meaning and propriety, even, of producing works of art and culture in colour, can also be found at the heart of the history of television's move into colour. It is found, for example, in the policy documents which surround the setting up of the first colour television service in Britain, on BBC2, in 1967; it is also found in the attitude of British broadcasters to colour broadcasts elsewhere in the world, as is shown in their public and private documents. However, one cannot underestimate the impact of colour on television screens, its vividness and luminosity, and the contrast that must have been so startling when television shifted from black and white to colour, first tentatively, on BBC2, in the summer of 1967, and then, by the end of the year, across the majority of the channel's programmes, and two years later, across BBC1 and ITV. The rhetoric surrounding the coming of colour, in the promotional material of the BBC and in reviews, editorials and letters from the public, speaks loudly of this contrast, and will be explored in this examination of the coming of colour to British television. The chapter will concentrate primarily on how colour television was conceived of before and during its initial appearance on British television, meaning that the focus will be on the BBC2 period alone. In this story of colour's arrival, particular attention is paid to both the internal discussions about colour technology and its capabilities within the BBC, but also to the ways in which this discourse was interpreted and expanded upon within the *marketing* of colour television.

Whilst others have given a more detailed technical history of the coming of colour, placing the 'race for colour' in its properly international context,[1] this chapter sketches the key points of this technical history only briefly, and offers an international history only in the sense that this leads to the question of the *tone and style* of the colour television that the BBC was seeking to produce for Britain (particularly as opposed to that which had been produced in North America). Thus this chapter shows that a 'restrained' and genteel colour palette, as opposed to the 'brashness' and 'gaudiness' of American colour, was written into television policy in the mid-1960s; it also looks at how this was interpreted as a search for authentic, realist or 'natural' colour by BBC management and programme makers, as opposed to a more 'spectacular' version of television colour. By analysing programmes, production files and listings guides that document the creation and broadcast of early colour programmes on British television, the success, or otherwise, of producing what might be seen as a chromophobic form of colour television in this early period can be assessed. In order to test these ideas, the impact that the coming of colour had on the production of television drama in particular is explored. It is clear that colour was assessed as having specific implications for drama producers, directors and designers: whereas, on the one hand, colour had the potential to add new layers of meaning to a production of television fiction and could therefore be used in expressive, and eloquent, ways, on the other hand, it might also be seen as producing a more superficial 'prettifying' of the television drama, offering an increased sense of spectacle and an attention to the *surface* of the image rather than producing greater depth of meaning.

The recent proliferation of writing on the role of colour in film has all begun by acknowledging that colour has been, until recently at least, an under-researched and misunderstood area of film scholarship. For example, Sarah Street has argued that 'For much of its history, commentators played down colour's contribution to film, emphasizing instead its function in underscoring dramatic narrative trajectories, neither drawing attention to itself nor acting as a distraction' (2010: 379). Three years earlier, Wendy Everett had been even more strident in claiming that film studies had a colour 'blind spot' and that within this field of scholarship 'there is no suggestion that colour might play any role in the construction of meaning' (2007b: 13). Whilst this must surely be seen as an overstatement, it is true that colour is, on the one hand, difficult to talk about in relation to film, and, on the other, is often understood as being of lesser importance than other aspects of mise-en-scène within a good deal of film analysis. This sense of *difficulty*, which also arises as an issue in art criticism, rests on the fact that, as is well documented, the perception of colour is subjective and culturally specific. For example, the naming of colour has been highlighted as a major issue for the film and fine art scholar alike, and reference to the Maori's '3,000 colour terms' (Everett, 2007b: 13) as opposed to the Filipino Hanunóo people's four has become a critical commonplace in representing this particular 'difficulty'. As Everett argues:

Spectacular colour?

> Whereas it may be tempting to think of certain colours as possessing universal meanings that are reflected in their use as symbols and codes, it is important to bear in mind that any relationship between colour and meaning is essentially arbitrary.
>
> (ibid.)

It is also true that even within cultures, we cannot be sure that colour perception is uniform. Therefore, scholars might be seen to shy away from using colour terms with authority and struggle with reading and understanding the *meaning* of colour, given the contingent and ambiguous nature of colour as a signifier.

The other major issue for film scholars has been the stability of film's colour. In short, if colour varies in reproduction from print to print, copy to copy, cinema to cinema, how can one write with certainty and conviction about it? Of course, if this variability is a problem for the study of film, it is infinitely more difficult and problematic when it comes to television. From the very first colour television sets, the 'look' of the image could be adjusted individually, 'to suit individual tastes', and sometimes in rather outlandish ways. Many sets allowed viewers to adjust brightness, contrast or tint and colour settings, and all sets had to be set up by a trained engineer who would adjust the colour reception internally before leaving the viewers to their own adjustments.[2] Furthermore, variation in the broadcast signal (often caused by viewers' distance from their nearest transmitter) had an impact on the stability of television colour, as did the transfer of programmes between different colour systems (i.e. NTSC to PAL). This issue of colour's variability was remarked upon in the reviews for the very beginning of BBC2's colour broadcasts, and the coverage of Wimbledon which kicked off colour in the UK, when Maurice Wiggin in the *Sunday Times* remarked that whilst 'we can rely on the BBC to put out good, lifelike colour [...] what we see at home will depend on things beyond their control' (Wiggin, 1967). Indeed, commenting on the Wimbledon coverage, the *Sunday Mirror*'s reviewer said,

> What a time I had fighting those two knobs marked Tint and Colour. At first the Centre Court looked like a sick greengage jelly in a sea of blue haze – which was the stands. Poor old David Attenborough, who introduced the programme, seemed

> to have chocolate blancmange all over his face. I fiddled and fiddled desperately with those knobs. And at last I got the trick of it.
>
> (Anon., 1967b)

A few days later, in the *Guardian*, Stanley Reynolds wrote that

> You may adjust the colour if you fancy surrealistic blues or greens or purples or if you wish to make some favourite television personality appear ridiculous. On BBC1 and ITV you can get a sort of sepia colour by twiddling the colour knob. The colour on our 25 inch set is very natural, more like real life than Technicolor, and with none of the green blur some have complained of.
>
> (Reynolds, 1967)

Clearly, then, reviewers of television were attuned to the variability of colour reception, and the fun that could be had in playing with television's colour – in essence 'mucking about with' colour in order to draw attention to the colour of the image and perhaps to render it more spectacular. The BBC was also worrying about this issue in their assessment of their own programmes; the Weekly Review of Programme Presentation minutes from 28 November 1967 documents a discussion with Barry Learoyd, the Co-ordinator of Colour Familiarisation, who wondered whether an area of greenness during the programme *Three of a Kind* (BBC, 1967) occurred 'because he lived on the fringe of the receiving area for colour television [and] if this was the case then he felt that it should be borne in mind when producing for colour since a major part of the general public would be viewing from a fringe area.'[3] What all this suggests is that like scholars of film who are concerned about the trueness of the colour they analyse in the particular print, video, DVD or download they are working with, television historians must, in dealing with colour, acknowledge that they are working with a *version* of a reproduction of television colour, rather than a guaranteed facsimile of what the viewer would have seen or what the director intended them to see. Nevertheless, with these caveats, it is possible to describe, analyse and evaluate the different uses of colour seen at the start of the colour television service in Britain; indeed, in all likelihood, the historian will be dealing with a version of television colour which is closer to the intended colours of the programme as broadcast than the colour that was received in homes across the UK in 1967 and 1968.

Setting up colour television and selling it

In brief, the facts of UK television's launch of colour are that the BBC launched a limited colour service in July 1967 on BBC2 only, which was already being delivered in the higher definition 625 lines system. After many years of experimentation and discussion (the Colour Policy files at the BBC Written Archives Centre go as far back as the start of the service in 1936), the UK adopted the German PAL system invented by Dr Walter Bruch of the Telefunken Company. Colour television had, however, begun in the US in the previous decade; CBS had introduced colour broadcasts as early as 1951, and whilst their colour broadcasting system was quite successful in technical terms, it was not adopted by other networks and was abandoned later that same year. NBC was then the first network to adopt the alternative NTSC colour standard, named for the National Television Standards Committee, in 1954. However, the take up of NTSC colour by other networks happened slowly across the next decade, with CBS, ironically, being the last to adopt the system in 1966, just a year before the BBC launched their colour service. Colour broadcasting in the US had had a shaky start: the NTSC system, referred to in the press and in the industry as an acronym for the phrase 'Never Twice the Same Colour', was seen as unreliable, and as producing gaudy, unnatural-looking colour. David Attenborough, who was Controller of BBC2 in the colour launching period, refers to the 'staggeringly garish quality of the first colour programmes shown in the United States' in his memoirs, and argues that 'I was sure, from watching the test transmissions, that ours would be in a different class, full of tonal subtleties and wholly comparable from the point of view of colour reproduction, with any printed colour pictures' (Attenborough, 2002: 212).

Attenborough's memories of the significance of this comparison are wholly accurate: a great deal is made in the Colour Policy documentation from the launch in 1967 about 'not doing it like the Americans'. The initial launching of the colour service in the UK was comparatively low-key, but with a larger number of colour programmes spread across the week, many of which were quite 'unspectacular' in their design; for example, *Late Night Line Up* (BBC2, 1964–72), a panel discussion/arts programme made in Studio H, the first fully equipped colour studio, was the first daily series to be broadcast in colour, and became something of a colour testing ground and 'flagship', despite its apparent lack of spectacle. In the month before colour was launched in the UK, in their document entitled 'The Colour Service', the BBC General Advisory Council reported on this quandary of whether to start with a 'few glittering colourful hours a week, set in a basic monochrome schedule' or produce the 'maximum amount of colour from the available equipment' (BBC General Advisory Council, 1967). This document noted that the Americans had opted for the former strategy and thus

> Producers and engineers [...] were faced with having to learn their skills on productions which would have been complex in monochrome but which in colour caused gigantic problems. Colour television thus became synonymous with complexity and difficulty [in the States, with engineers becoming] so obsessed with colour that it frequently came to dominate their production plans. As a result the crucial programme values – the wit of a comedian, the dramatic quality of a play, the balance of news coverage – became of secondary importance.
>
> (ibid.)

In comparison, the General Advisory Council argued,

> Our aim will be to produce not a few isolated colour programmes, but a complete colour service [...] Thus, programmes of all kinds – discussions as well as operas, science documentaries as well as light entertainment spectaculars – will be produced to this new standard. There will be no question of putting only 'colourful' productions into a colour studio or of striving to add pretty colours to a programme where they are irrelevant or inappropriate.
>
> (ibid.)

The BBC thus strove to produce colour television with a 'more subtle' tone and style than the Americans had done.

A crucial difference between film and television colour is that as soon as a film is made in colour, everyone sees it in colour, whereas the shift into television colour required financial investment on the part of the viewer, in the form of purchasing a new colour set and investing in a TV licence, which was £5 more expensive than a regular licence per year. For those still watching in black and white, colour television could be viewed as black and white, so the reception of television in colour was far more gradual than that of film. According to Andrew Crisell, a colour set would have cost £350 in 1967 (the equivalent of around £5,000 in 2012 when adjusted for inflation) (2002: 122). This was, therefore, no small investment, and accounts for the fact that only 17 per cent of households were equipped with colour TV by 1972 (Lury, 2005: 36), though this number rose rapidly throughout the 1970s, thanks largely to the hiring and hire-purchasing of colour sets through companies like Radio Rentals and Rediffusion. Whilst Radio Rentals had ensured that the TV critics were viewing in colour by sending them free colour sets,[4] Maurice Wiggin reported in the *Sunday Times* in the first week of colour that there was only an estimated one to two thousand sets in operation across the country.[5] Thus the BBC found itself at the forefront of promoting the purchase of colour television, and worked closely with BREMA, the British Radio Equipment Manufacturers' Association, to ensure that there were enough viewers purchasing new sets, and therefore the new, more expensive, licence, to cover the increased costs of moving to 625-line colour programme production and broadcasting.

The jewel in the crown of the BBC and BREMA's joint efforts was the travelling exhibition, *Colour Comes to Town*, which is described by the BBC's Publicity Service as being 'comprised of 14 manufacturer's stands and four BBC stands [...] a total of 308,000 people visited the Exhibition at its eight venues'.[6] The exhibition opened at the Fairfield Halls in Croydon on 25 September 1967, and the BBC employed three 'Colour Girls' – actresses/singers from London, Skipton and Glasgow – to host the exhibition in the relevant regional centres, answering visitors' questions, introducing celebrity guests from the world of television and keeping an eye on the mobile colour TV demonstration unit which was displaying colour showreels at the exhibition, and a back projection unit showing a continuous selection of slides illustrating colour programmes. For many people without the capital, and without rich enough friends and neighbours, *Colour Comes to Town* was their first experience of colour television, just as the large-scale mid-century exhibitions had introduced visitors to television for the first time. Following the buzz (and high attendance figures) of this mobile exhibition, the BBC and BREMA sent a scaled-down version of *Colour Comes to Town* round to tour holiday camps the following summer. The BBC Board of Management minutes for 22 July, 1968, report that this venture had been 'a considerable success [and that] It appeared that 90% of those visiting the exhibition had never seen a colour television picture'.[7] The same Board reported in the following month that

> at the end of that week [the holiday camp tour] would probably have reached 25,000 people […] [and] at one camp about 150 people had sat through a performance of *Cosi Fan Tutte* [being broadcast on a colour set in the exhibition].[8]

Alongside BREMA, the BBC had also mounted an exhibition in the fifth-floor exhibition lounge of Austin Reed's in Regent Street during Wimbledon Fortnight in 1968, and, teaming up with Kodak, had simultaneously mounted an exhibition entitled *Colour Comes to Television* in Harrods' 'Fashion Theatre'.[9] Thus the BBC aimed to cover all of their demographic bases in collaborating with set and film stock manufacturers, as well as TV rental companies, to promote colour television within a variety of public spaces during the launch period. The 'show business' tone of this promotion, and its emphasis on the spectacle of colour TV, are interesting because they sit slightly at odds with David Attenborough's insistence that the colour launch would *not* be spectacular, but rather should be seen as the *natural* progression of television form and style, and as an increase in fidelity rather than a recourse to showier aspects of the nascent colour service. It is this quandary to which we now turn.

On the question of 'spectacular colour' vs 'realist colour'

Francis McLean, the BBC's Director of Engineering, opened his lunchtime lecture in March 1967 by asking why the BBC was switching to colour. In answering his own question, McLean said:

Spectacular colour?

> Well, the Government has told us to do it, and of course we wanted to do it for various reasons, like keeping up with the Joneses, helping the export trade, and so on. These are all important, but I think the real reason is that it is the natural thing to do. By a sort of curious inversion of logic, we have come to regard black and white television as a normal thing, and colour as abnormal, whereas in life the natural thing is to see everything in colour.
>
> (McLean, 1967: 3)

McLean went on to say that

> It is sometimes said that colour is an extravagance and that we could do without it. This is of course true – we can do without a great many things [...] but experience shows that human nature does not work this way. Colour is such an essential part of human living that it would be an unnatural deprivation to do without it.
>
> (ibid.: 15–16)

The rhetoric used here is indicative of the BBC's drive to produce 'natural' colour, to see it as an essential, and not unnecessary, progression of the television image, as something that would augment the BBC's 'core values' of providing truth and extending cultural experience, rather than something which would detract from this. In short, the coming of colour was, according to management-speak at least, to be precisely un-televisual, using John Caldwell's (1995) understanding of that term to mean 'excessively styled' or exhibitionist in its aesthetic. A month after McLean's lecture, the trade publication *Electrical and Radio Trading* reported that 'the very word colour, Mr Attenborough said, was a misnomer. It puts a false emphasis. We should say natural television. Monochrome has been an impoverished and debased picture. We shall now provide a more accurate picture' (Anon., 1967a).

BBC management went to great lengths to deny colour television's more spectacular functions or properties during the launch period, as reflected in Director General Charles Curran's statement that colour should not 'simply [be] a decorative addition to the screen' (1969: 7). Whereas the term 'high fidelity' colour often replaced 'realist' colour within BBC policy documentation, the meaning of this was certain. The term *faithfulness* was also used to denote a realist approach to colour: the Controller of Programming for Television, Ian Atkins, argued that 'faithfulness to the original is obviously the ultimate criterion of colour pictures' and adamantly stated that 'This new dimension that has just been added to the television picture will not be exploited for its own sake'.[10] This rhetoric runs contra to the histories of British television which argue, as Crisell does, that the 'obvious effect of colour was to make the medium of television immensely more vivid and picturesque' (2002: 122). Furthermore, when one looks at the reception of what Karen Lury describes as a 'homegrown spectacular' (2005: 36), that is, the BBC's colour coverage of Wimbledon in 1967, what becomes clear is that the reviews of this suggest it was both eye-poppingly spectacular and at the same time stress the realism of the colour coverage (the grass is actually green, the tennis whites, white and so on). In Attenborough's summing up of these reviews in an editorial for the *Radio Times*, the superlatives used ('Marvellous, awesome, true-to-life, epoch-making, a new dimension'; Attenborough, 1967: 5) speak simultaneously to colour's potential to be both spectacular (marvellous, awesome, epoch-making), but also closer to the real (true-to-life, a new dimension).

However, it was variety programming, even more so than sport, that was seen as one of the key beneficiaries of the shift into colour, relying, as the genre did, on spectacular sets, costumes, lighting and choreographed set pieces. Furthermore, it was the variety programme which took the BBC closer to the garish American colour palette which Attenborough et al. had been so keen to avoid, seen so vividly in the acid-bright, primary colours of sets, lighting and costumes of the *Black and White Minstrel Show*. In his report, 'Colour Television: a Report on Progress', Secretary to the BBC Kenneth Lamb states that 'The only concession to colour', in terms of a shift in programme policy, was the transfer of the highly popular variety show, *The Black and White Minstrel Show*, from BBC1 to BBC2. He argues that 'This was necessary because BBC2 had no light entertainment "spectacular" through which to demonstrate the potential of colour when used in productions of this kind'.[11] However, this is not strictly the case, given that BBC2 launched its full colour service with 1 hour and 15 minutes of *Billy Smart's Circus* (BBC2, tx. 2/12/67), a spectacular variety event programme, and, in the first weeks of the full colour service, broadcast either the *Black and White Minstrel Show* (BBC1/2, 1958–78) or variety/comedy vehicle, *The Charlie Drake Show* (BBC2, 1967–8), on Saturday evening, and *Once More with Felix* (BBC2, 1967–70) on a Sunday evening. Whilst the latter was based around the folk singer Julie Felix, it still had the format and design of a variety show, and its innovative use of coloured lighting was praised highly by the Head of Television Lighting, Phil Ward.[12] Of the *Charlie Drake Show*, the producer Ernest Maxim said in the *Radio Times* that

> The word 'spectacular' sounds as though we're letting off fireworks. In a way, it's true [...] It will certainly be colourful. The costumes are marvellous and we have lit the sets like a Hollywood film. A black and white show, when completed, can give certain satisfaction, but a colour show 500 per cent more.
> (Anon., 1967c: 11)

The colour design and prominence of these BBC2 variety shows in the schedule, also referred to as 'Spectaculars' within policy documents produced in the 1960s, sits at odds then with Attenborough et al.'s suggestion that BBC2 would launch a more subtle colour service than the Americans that would not seek to 'showcase' this new technology with 'excessive' production values but rather to use colour as a 'natural progression' towards a greater sense of verisimilitude. Whilst it is clear to see that the colour spectacular did not lend itself to 'natural' or 'subtle' presentations of colour, the extent to which this was achieved in television drama production is explored below.

Making meaning or 'looking pretty'

In considering whether some dramas should continue to be produced in black and white after the launch of the colour service, David Attenborough drafted a letter to Kenneth Adam, Director of Television, in 1966, with the following thoughts: 'There may be strong aesthetic reasons why some productions should be shown in monochrome. This will doubtless apply to many plays, for colour, unless used with great skill, can turn the dramatic into the merely pretty.'[13] Whilst Attenborough changed his 'many plays' to 'some plays' in the final draft of this letter, the fear of producing the 'merely pretty' television drama remained plain. In her writing on the 'pretty' in cinema, Rosalind Galt argues that 'prettiness' is frequently at the centre of 'the resilience of "empty spectacle" as a figure of critique in film writing from journalism to high theory' (2011: 2) and goes on to say that

> Pretty things do not have the status of beautiful ones [...] because pretty so immediately brings to mind a negative, even repugnant, version of aesthetic value [...] Many critics hear in the term a silent 'merely' in which the merely pretty is understood as a pleasing surface for an unsophisticated audience, lacking in depth, seriousness, or complexity of meaning.
>
> (ibid.: 6)

Galt's etymology of 'pretty', and the 'merely pretty' more specifically, is useful in understanding precisely the formal elements Attenborough feared colour might bring to television drama: the production of decorative or attractive drama which lacked depth, seriousness and complexity, an emphasis on (tele)visual pleasure over the construction of meaningful narratives.

The first long-running colour dramas on BBC2 during the colour launching period were American imports: these included the anthology drama *Impact*,[14] which featured high-end actors like Simone Signoret, John Cassavetes, Dana Andrews and Shelley Winters, and aimed to bring the values of Hollywood colour cinema to the small screen, and the westerns *The Virginian* (NBC, 1962–71) and later *High Chaparral* (NBC, 1967–71). In an article entitled 'My Verdict on Colour TV', Shaun Usher in the *Daily Sketch* described *The Virginian* as 'the dullest, wordiest, most pompous Western series ever screened on television' (Usher, 1967), but then argued that, in relation to its broadcast in colour, 'all of a sudden I wouldn't be without it for the world [...] For whenever the action palls, my eyes wander to the scenery – green pine trees, vivid blue or sunset skies' (ibid.). We see here an account of a show doing exactly what Attenborough had feared of colour drama: lacking in dramatic value but providing a rather more 'shallow' sense of visual pleasure in colour not tied to the creation of meaning. This review, then, sees colour landscapes being viewed as salve for poor drama, rather than, as Technicolor pioneer Natalie Kalmus proposed, colour being utilised to 'convey dramatic moods and impressions to the audience, making them more receptive to whatever emotional effect the scenes, action, and dialog may convey' (2006: 26).

In preparing for the start of the colour service, the BBC's own producers, directors and designers of television drama, along with all other technical staff, needed to be trained in the specific issues and problems of producing TV fiction in colour. It was decided that this training had to be done in such a way that the makers of television drama (along with all other genres of programming) would produce full programmes, as if for broadcast, to test out the new equipment. This training, dubbed the 'Colour Familiarisation Course', was described by its co-ordinator, Barry Learoyd, in a memo to director Rudolph Cartier's assistant:

> There are precisely three weeks from [the start date] during which the production team must prepare, plan, design, rehearse and, in the studio, bring to a successful final run your team's 20–30 minute programme [...] The object of the course is for you to obtain as much practical experience of colour and its compatible black and white picture as possible, in the time [...] Simple programmes with a simple use of colour are likely to be more successful until the Service as whole has gained experience [...] Colour is accentuated on the screen. Pictures can easily become 'gaudy'. Most successful pictures are composed largely of subdued colours.[15]

Ian Atkins later said of this course that the working teams 'were encouraged to experiment and if they got it right first time we asked them to try it another way. We believed that they would learn as much from their failures as from their successes'.[16]

The training took place in Studio H and Cartier's team produced a production called *Three Essays in Colour* featuring three literary adaptations (Chekov's *The Bear*, the fantasy mystery novel *The Master of Judgement Day* by Leo Perutz, and Akutagawa's *Rashomon*) and experimented with, amongst other things, the use of back projection in colour production and also highlighted the major problem of shifting between studio and footage filmed on location: this had always been seen as a problem for the production of dramas, but even more so in colour. From the documents pertaining to the colour familiarisation course to the files of ongoing series and serials once colour production began in earnest, there are repeated issues about the disjuncture between these elements of productions: actors' hair appearing to be two different colours in the same scene, weather and lighting changing starkly from shot to shot, etc. Coming out of this process, there was a repeated insistence within the BBC on an increased need for teamwork and *collaboration* among the directors, producers, designers and costume and make-up designers of TV drama; as Kenneth Lamb argued, 'colour demands that all the separate creative processes of production should be considered in relation to each other'.[17] This is also evidenced in a document written by Head of Classic Serials, David Conroy,[18] which sets out a five-step process of collaboration in a series of 'Colour Co-ordination Meetings' between the director and all other members of the design and realisation team on each programme. Given that colour programming was around 20 per cent more expensive to produce than black and white,[19] and that the differential costs for the classic serial were slightly higher at 23 per cent, Conroy was wise in carefully laying out the processes of colour production for all who worked on the classic serial, for it was the classic serial which was to launch the BBC Drama Department's efforts in colour.

The BBC launched their full colour service on BBC2, with a five-part adaptation of *Vanity Fair,* adapted by Rex Tucker and directed by David Giles. The adaptation was strongly promoted: stories about its production appeared in a weekly colour supplement of the *Radio Times* in the five weeks running up to the broadcast of the first episode, and its star, Susan Hampshire, appeared on the front of the *Radio Times* in the first week of the full colour service, advertising both *Vanity Fair* but also the start of colour. Following this, and broadcast in the same Saturday and Thursday evening slots, Jack Pullman's adaptation of *Portrait of a Lady*, directed by James Cellan Jones, was produced. In their history of the classic serial on television, Giddings and Selby argue that 'Colour television [...] brought a revival of interest in classic novels which could be set in luscious locations, classy architecture or rhapsodic landscapes' (2001: 30). The costume drama, and the classic adaptation in particular, is, of course, one of the few areas of television production to have previously been considered in relation to questions of visual pleasure. Whilst analyses of the genre have mainly concentrated on the sumptuousness of costume dramas shot on film and on location from the 1980s onwards, I have previously argued that the studio-based costume drama might also be understood in relation to similar notions of visual pleasure, and particularly the pleasures of heritage detail found in costuming and set dressing (Wheatley, 2005). This is certainly the case for both *Vanity Fair* and *Portrait of a Lady*, which were both shot mainly in the studio, with some location-filmed inserts for exterior shooting: their costuming and set design fit Kristen Thompson's description of colour in film as described by Brian Price: 'excess[ive], a source of abstract visual pleasure and perceptual play that resides somewhere above' questions of meaning (Price, 2006: 6). *Vanity Fair* was commissioned by Shaun Sutton, Head of Serials in the BBC Drama Department, because:

> (1) It is a Classic, but not too heavy or demanding a Classic [...]
> (2) It has a splendid girl as its leading character. A girl who is wicked, but not too wicked and (3) It is good colour. A cool, elegant period in architecture, very pretty women's dresses, attractive men's clothes and uniforms.[20]

Whilst Sutton tempered these comments, found in a memo to the Controller of Television, by following them with 'Of course I agree it is the content, and only the content, that matters in the end. Scenery in television is the thing that stops you seeing the studio walls',[21] his initial emphasis on *Vanity Fair* being an 'unchallenging' play that offered the opportunity for 'good colour' is telling.[22]

Vanity Fair is a sprawling tale of the Machiavellian manoeuvrings of Becky Sharp, a young woman intent on social climbing who befriends and betrays the family of her kind and innocent school friend, Amelia Sedley, amongst others, in her pursuit of fortune and social esteem. Sharp, played by Susan Hampshire, is an interesting figurehead for colour television: a vain and deceitful figure, she is repeatedly shown to be obsessed with the appearance of things. Beautiful in comparison to her much plainer friend, Amelia, Hampshire plays Becky as a coquettish but fiery red-head (despite Thackeray's description of Becky as 'sandy haired'), perhaps literally embodying what colour technology was capable of showing. This reading of Becky as an embodiment of colour is also evidenced by the fact that Becky's costumes are an array of colours – pinks and purples and greens, – as opposed to the whites and pale blues of Amelia's. When approaching a colour-centric reading of *Vanity Fair* (or any other piece of television, for that matter), we seek out some evidence of colour being used in meaningful or expressive ways, looking for what colour brings to the creation of meaning in this production or lent to the symbolic aspects of storytelling. In the contrast between Becky as a kind of peacock, an embodiment of Batchelor's 'chromophobia' in which the colourful is also the deceitful, dangerous and shallow, as opposed to the purity, innocence and truthfulness of Amelia, we do see this symbolic use of colour, to a certain extent.

However, what is far more striking about the use of colour in *Vanity Fair* is that it is frequently figured as an enhancement of the decorative elements of the mise-en-scène of the classic serial: this novel, chosen for adaptation for its series of balls and gatherings of men in colourful military uniform, offers the programme makers ample opportunity to *showcase* colour, with a parade of contrasting costumes running through the entire colour spectrum. Short montage sequences, such as the moment in the first episode when a number of characters prepare to attend a party in Vauxhall, present a spectrum of colour: the characters, dressed in a range of colours from the deep purple velvet of George's jacket to the bright red of Captain Dobbin's military dress, and, of course, Becky's sugary-pink gown, are shot in medium close-up, facing the camera, which briefly lingers on the richness of these fabrics. There then follows a filmed insert of a night sky bursting with fireworks, and then a tracking shot around the Vauxhall party, with coloured streamers hanging down all around the group as they arrive. Here, as in many of the dance and party sequences of *Vanity Fair*, we see colour being strikingly used for its *decorative* properties. An array of colour is offered but without any obvious sense of symbolic meaning: it is colour to be looked at, to be enjoyed, rather than to be understood. The colour design reflects the gaiety of the sequence and the characters' lives at this moment, but more than that, it expresses a joyousness in colour itself. As Becky dances in circles, filled with excitement and enacting this sense of deep joy, we see a moment, like many moments in *Vanity Fair*, when the pleasure of colour is foregrounded – perhaps not entirely 'merely pretty' drama, but in terms of the definitions set out in the colour policy discussed in this chapter, perilously close to it. This sense of the pleasure to be taken in colour which is emphasised by *Vanity Fair* is reflected in a

Spectacular colour?

viewer's letter to the *Radio Times* from a Mr Cecil Williams of Wendover, Buckinghamshire:

> I cannot praise the BBC too highly for the quality of their colour programmes. *Vanity Fair* took on a new beauty. The elegant richness of the costumes and the detail of the settings were revealed as in a new dimension [...] Having first had TV in 1950 when it was fairly rare, we have seen many and varied programmes and have become selective. This colour is something which exceeds our expectations in every way.
>
> (Williams, 1968)

It is notable that this 'early adopter' uses the term 'beauty' rather than 'prettiness' in describing *Vanity Fair*, remembering that Galt (2011) argues that the beautiful is more culturally acceptable than the merely pretty.

The Owl Service was the first Granada drama to be shot entirely on colour film and on location in the following year. If we compare its use of colour with *Vanity Fair*'s, for example, we see a strikingly different approach[23]. Jealousy, class struggle, burgeoning sexual desire and suggestions of incest are writ large through each episode of *The Owl Service*: the action centres around the developing triangular relationship between Alison and Roger, step siblings, and Gwyn, the son of the housekeeper of the house in the Welsh valleys where Alison and Roger have been brought for the summer. What is particularly striking about this adaptation though is that the 'love triangle' between the three central characters is defined via a symbolic use of colour in the drama: the colour scheme, based around the 1960s wiring system, sees Alison, a sexually charged young adult whose body is the central object of desire, dressed entirely in 'live' red throughout, Gwyn clad in black to suggest his relationship to the 'earth', and his groundedness in Wales and the Welsh landscape, and the jealous Roger dressed only in 'neutral' green as an expression of his frustration and jealousy. Here colour is deeply meaningful (almost hysterically so), rather than 'merely pretty'.

In conclusion, it must be pointed out that the classic serial wasn't the only genre of drama to be made in these early days of colour at the BBC; as discussed in Wheatley (2006), an anthology series called *Late Night Horror* (BBC2, 1968) was the first series to be produced in colour by Harry Moore and the directors who had done the colour familiarisation course, including Rudolph Cartier and Naomi Capon, though wasn't broadcast until 1968. The production files for *Late Night Horror* show that Moore was keen that his directors exploit the new possibilities of colour to the full, with lots of blood and guts on show (as discussed in Wheatley, ibid.). Furthermore, the long-running anthology series *Theatre 625* (BBC2, 1964–8) and *Thirty Minute Theatre* (BBC2, 1965–73) also began producing in colour towards the end of 1967; as anthology dramas on which a variety of teams worked, they offer a fascinating prospect for the historian researching the impact of colour on drama production. How the teams working on individual episodes responded to and used the new possibilities of the medium might offer us an illuminating and varied picture of colour's dramatic possibilities, and Leah Panos' (2015) forthcoming work looks precisely at this history, and, in particular, the episodes of these two series directed by Rudolph Cartier. Beyond Cartier's work on these series, few episodes remain, unfortunately. It would be fascinating, for example, to see the episode of *Thirty-Minute Theatre* 'Lovely in Black' from 24 January 1968, in which a young woman imagines people visiting her after the death of her husband (who, it is revealed at the end of the drama, is not actually dead). A production note in the file for this episode states that

> Olivia's imaginary life, being more colourful than reality, takes place in an atmosphere almost gay, despite Alfred's supposed demise. Only at the beginning and end of the play is there the greyness imposed by loneliness and Alfred's personality. The quality of light can be used to suggest this, and the colour of Alfred's clothes.[24]

It is important to remember, ultimately, that these were dramas that were still being made for the majority of the audience to view in black and white: as David Conroy, Head of Classic Serials, stated in a press release for *Vanity Fair*, 'This first colour drama contribution to the full colour service will still be "colourful" in black and white. As much care is being taken in monochrome tonal values as is being taken in colour aspects'.[25] Perhaps this explains the lack of depth and complexity in the use of colour in the early costume dramas discussed? Certainly, this raises an interesting historiographical issue: whilst our instincts in researching the impact of colour technology on television drama might be to look at the very first

Spectacular colour?

productions made in colour, perhaps we might see a more developed use of colour in programmes made and broadcast at the end of the following decade, when most viewers were watching in colour and when directors and designers might be fairly sure that most people would be able to *read* the colour of their productions? Even if, as argued at the start of this chapter, the meanings of colour can be understood as subjective, even arbitrary, it is clear that the makers of television were attempting to use colour in both spectacular and meaningful ways, as well as bringing a new kind of visual pleasure to television, and television drama in particular, in these very early days.

Part II

Spectacular Landscapes and the Natural World: Exploring Beautiful Television

3

At home on safari: Colonial spectacle, domestic space and 1950s television

In her book *Imperial Eyes: Travel Writing and Transculturation*, Mary Louise Pratt describes what she calls the 'contact zone' of the colonial world (1992: 8), urging us to examine the spaces and moments of contact and collision between cultures, rather than accepting their inherent separateness, and to look at the ways in which both coloniser and colonised are transformed within this contact zone. Sara Mills' work on women's travel writing has also drawn on this notion of a contact zone in her examination of the meanings and organisation of domestic space within the colonial context; she argues that these domestic spaces provided significant 'contact zones' between Europeans and, particularly, European domesticity and the people and culture of the colonised nation (2005).[1] In 1950s Britain, looking out from the metropole to the colonies, television also became an important 'contact zone' between cultures at a crucial point in the history of the British colonies, during their struggle for independence. Taking the safari series of the popular husband and wife presenting team Armand and Michaela Denis as its central example (specifically their programmes *Filming Wild Animals* (1954–5) and *On Safari* (1957–65), produced for the BBC in Kenya),[2] this chapter illuminates the ways in which television might be understood as a colonial contact zone in the 1950s. By paying particular attention to the looking relations, sound design and landscape photography of this programming, we come to understand the part played by spectacular television in forming a picture of the Kenyan people and their landscape under colonial rule. As Jeffrey Richards has noted in his discussion of cinema and imperialism in the 1930s,

> For ordinary people, the Empire was the mythic landscape of romance and adventure [...] most people were not bothered about the actual conditions in the Empire. It was the imagery which they absorbed and endorsed and that imagery was romantic, adventurous and exotic.
>
> (1986: 144)

Whilst there is a robust, and growing, body of critical literature on colonial cinema on which this chapter will selectively draw, the question of colonial television is yet to be explored thoroughly.[3] This chapter thus aims to open up this debate in the context of Kenya and Britain in the 1950s, specifically in relation to wildlife programming and its representation of the flora, fauna *and people* of Kenya. There are, of course, broader implications in establishing an understanding of the term 'colonial television' beyond the specific history of the nature, travel, historical or anthropological documentary programmes, or indeed the action-adventure series set in the colonies, during this pivotal period in colonial history. Television itself has been understood as a 'colonising' apparatus, and metaphors of the colonial have become rooted in our understanding of television as a medium of spectacle.

Colonial Kenya

The story of how land, landscape and wildlife figured in the history of colonial Kenya is an important starting point.[4] In 1887, the British East Africa Association was granted a concession by the Sultan of Zanzibar to full judicial and political authority over his mainland possessions from the Umba river in the south to Kipini in the north, in return for paying him a proportion of the custom dues they were to collect. Within a year of its inauguration, the Association claimed rights for 200 miles inland and in 1888 Sir William Mackinnon and the Imperial British East Africa Company (IBEAC) were authorised by the British government to develop and administer the 'eastern territory'. In effect, in these early years, Kenya, known at the time as part of the East African Protectorate (EAP), was an obstacle for the IBEAC, an expanse that had to be crossed between the coast and their territory inland in Uganda. When the British government took over the East African region from the IBEAC in July 1895, the IBEAC had not made sufficient funds from the region (early settlers had found life in the EAP treacherous, and farming and transport had both initially proven extremely difficult in the country); the full 'exploitation' of the area could, it was thought, only be achieved by the building of a railroad from Mombasa on the coast through the highlands and towards Uganda. On the boggy plains that the railroad eventually passed through, the city of Nairobi was established, and from the building of the Uganda Railroad, the relatively small numbers of white settlers began to increase. Its construction had been very expensive for the British government, and as David Fieldhouse has argued 'once it became clear that only large-scale European-type farming in the White Highlands could make the railway pay its costs, London had no hesitation in encouraging white settlers, even though this involved extensive eviction of the Maasai and other groups' (1996: 130). The East African Protectorate therefore soon became figured in the British national imagination as a land of opportunity: 'The British persuaded themselves in the late nineteenth century that it was full of wide-open, scarcely populated areas, a "new Australia" in the enthusiastic phraseology of the time' (Chamberlain, 1999: 49). Fighting on the German East African front during World War I had strengthened British settlers' sense of collectivity and *belonging* – white farmers joined up to the East African Mounted Rifles, and many resourceful women ran farms and homesteads whilst their husbands were away fighting. As C. S. Nicholls argues, victory reinforced white British self-confidence in the EAP and their belief in the invincibility of empire: settler numbers were buoyed by the raffle of farmland to demobbed servicemen (some of these plots formerly in African reserves and some divided from already alienated land), so that between 1914 and 1921, numbers of whites in the EAP, renamed Kenya in 1920, had risen from 5,438 to 9,651 (Nicholls, 2005). In the period before and immediately after World War I, therefore, there was what Richard West describes as a 'Wild West atmosphere' (1965: 15), a scrabble for the best highlands in which highly successful tea and coffee plantations could be set up. Whilst many people ended up abandoning their farms over the years to come, thwarted by a lack of farming experience, disease and attacks from wildlife, an invasion of locusts which decimated crops, and the international financial depression of 1929, Kenya continued to be figured in the popular British imagination as a land of opportunity.

In thinking about Kenya as it was figured in the Anglo-American imagination during the first half of the twentieth century, it is also important to acknowledge the centrality of the safari in this history.[5] During the period of Kenya's colonisation, the social character of hunting there shifted considerably: whilst Kenyans were increasingly barred from hunting on their own land, primarily through the establishment of game reserves, a massive hunting and safari industry grew up out of Nairobi, serving the wealthy white tourist. Safaris had been popularised in the first couple of decades of the century by high-profile hunting visitors. These included Winston Churchill in 1907, the Duke of Connaught, the honeymooning Duke and Duchess of York, and Theodore Roosevelt, as well as the cult of the 'white hunter', dashing safari guides working for companies like R. J. Cunninghame and A. C. Hoey that sprang up in the wake of these earlier 'celebrity' safaris. The impact on the Kenyan landscape of the safari industry was profound: with the conversion of game from source of food, ivory and hides to a means of raising revenue via sport and tourism, enormous game reserves were established all over Kenya. At the height of this practice, the country was figured as a kind of 'playground' for rich Americans and Europeans. For example, H. C. F. Wilkes, general manager of Ker, Downey and Selby Safaris Ltd. stated in the early 1960s that

> the ordinary American couple can spend more in a month on a safari – without blinking an eyelid – than I've ever earned in a year. The basic cost per person is £1,350,[6] plus licence, drinks, ammunition and film. You might even have some extra mileage. And, of course, this doesn't include the air fare.
>
> (West, 1965: 18)

Even in the mid-1940s when the National Parks were set up, partly to serve a different kind of safari which might be understood as 'hunting with cameras' rather than with guns,[7] 'live animals became a tourist resource, a vast outdoor zoo, their haunts made accessible by the building of roads, hotels and rest camps' (MacKenzie, 1988: 201). Arguably, many of the films and television programmes that depicted Kenya and Kenyan wildlife drew on the earlier popularity of the safari.[8]

As the drive to colonise what was presented as 'empty land' (Maugham-Brown, 1985: 23) gathered pace, the white British living in and writing about Kenya continued to assert the myth that this was wide-open landscape, sparsely populated by black Africans who welcomed their colonisers into this utopian setting. The ways in which Kenya was *imagined*, and the ways in which the Kenyan landscape entered into the vernacular, were to be particularly significant in the construction of space and place in colonial cinema in the interwar years and beyond, and would also become central to the wildlife programmes of Armand and Michaela Denis. Introducing the 'problem' of discussing landscape in the context of colonial and post-colonial Africa, environmental historians William Beinart and JoAnn McGregor have noted that,

> In the past, many historians of Africa were reluctant to use the term 'landscape' because they felt it to be too firmly associated with a particular European tradition of seeing, related to the emergence of specific property relations and artistic expression, and therefore inappropriate for other contexts.
>
> (2003:4)

However, Beinart and McGregor argue, 'by defining "landscape" broadly, as an imaginative construction of the environment, new areas have opened up for Africanists', bringing together histories of changes in the African environment with 'imaginative interpretations' of landscape (ibid.). They note, for example, that 'landscape ideas were [...] important in constructions of settler identity' and that 'settlers from diverse backgrounds could be invited to unite around a love of "wilderness" or an idealised rural background' (ibid.: 5). This 'love of the wilderness', and a fascination with African wildlife and the customs of African people, found their most profound expression in the films and television programmes of empire, and specifically in the spectacle of Kenya's people, wildlife and land.

There is, of course, an audio-visual history of the depiction of Africa (and Kenya, more specifically) to take into account when contextualising the Denises' programme-making that stretches beyond television, and there are numerous histories of Africa on film to which we can turn here in thinking about the production of colonial film in this context.[9] This history covers the early safari films of Cherry Kearton (e.g. *Theodore Roosevelt in Africa*, 1909) and Paul Rainey (e.g. *Rainey's African Hunt*, 1912), the relative explosion of amateur safari films following World War I, and the development of government-sponsored filmmaking in Kenya (via the Empire Marketing Board,[10] British Instructional Films[11] and the Colonial Film Unit,[12] for example). However, it is Martin and Osa Johnson, described by Kenneth Cameron as 'the world's first husband-and-wife, real-life-adventure film team' (1994: 123), who can be seen as the blueprint for the Denises. Their first film, *Trailing Wild Animals* (1923), like the Denises' programmes, was shot out of Nairobi and constructed from a vast collection of footage that they had amassed on their sorties into the bush. Whilst the British filmmakers Major and Stella Court-Treatt later tried to 'turn away from exoticism and condescension and to have tried to turn their cameras on a real – and serious – Africa' (Cameron, 1994: 50),[13] the Johnsons, like the Denises would 20 years later, combined the excitement of the safari with Osa's on-screen 'cutesiness' and established a way of seeing Kenya and presenting its landscape which later wildlife programming mirrored.

Armand Denis had also played a significant part in portraying Africa in the cinema, prior to his move into television. Denis was a Belgian inventor, naturalist and filmmaker, who made a number of feature-length melodramatic Hollywood safari documentaries such as *Wild Cargo* (1934),[14] *Savage Splendour* (1949)[15] and *Below the Sahara* (1953), all for RKO. Denis had begun filmmaking as cameraman to his future father-in-law, André Roosevelt, on the Balinese action adventure film *Goona Goona* (1928), which combined expedition footage with a story of a romance between a Balinese prince and a servant girl. He later married Roosevelt's daughter, Leila, with whom he had four children and with whom he continued to make films; *Dangerous Journey* (1944) is an account of their travels in Africa, India and Burma. In 1948, however, he met, had an affair with and married his second wife, Michaela, bringing her over to Africa initially to finance their future filmmaking by working together on *King Solomon's Mines* (MGM, 1950): Denis was acting as consultant on this shoot, whilst Michaela was brought in as Deborah Kerr's body double. After working on this feature, they made the film *Below the Sahara* together in 1953, which, coming just a year before their first appearance on the BBC, must have drawn them to the Corporation's attention.

Later, and contemporaneously with the Denises' work for the BBC, Kenya briefly became a focus of fictional feature films in the mid-1950s, films which were to represent the Mau Mau uprising as an extension of the danger and excitement of the safari film. Films such as the Dirk Bogarde vehicle *Simba* (1955) and *Safari* (Columbia, 1956), which had the tag-line 'Murderous Mau-Mau! Maddened Beasts! Mighty Jungle Love!', firmly placed Kenya in the 'Dark Continent' and depicted black Africans as untrustworthy at best, and killers at worst. Kenneth Cameron argued that:

> Mau Mau terrified Britons in both the colonies and England. It seemed to threaten whites with the very horrors that had dominated the imagery of Africa: whites overcome by the sheer numbers of blacks; whites hacked to pieces by blacks armed with primitive weapons; whites destroyed by a culture based in 'magic' [...] In actual fact, thirty two white colonials died as a result of Mau Mau [...] More than seven thousand blacks died. Yet the myth of Mau Mau was a myth of mass white death and so it came into popular white culture.
> (1994: 114)

Whilst many, including Cameron, place the 1955 Dirk Bogarde vehicle, *Simba*, within this tradition of representation and cite it as the ultimate example of racist, colonial filmmaking, Christine Geraghty carefully describes the contradictory nature of the politics of the representations in the film (which she locates in a broader history of the Commonwealth film). Geraghty believes that *Simba* combines 'the liberal discourse of the new Commonwealth within generic conventions that are strongly linked to the films of empire' (2000: 125). However, given that these films were directly contemporaneous with the Denises' wildlife programmes on British television, and also that, as Chapman and Cull argue, 'the pervading theme of British-made empire films of this period is the idea of the empire under threat' (2009: 8), the Denises' narratives of Kenya as a space of peace and natural beauty to be protected by 'white and black Africans' alike, and of the people of Kenya as both curious and cooperative (rather than threatening or murderous), are perhaps all the more striking.

As well as reflecting on colonial filmmaking in Kenya in the period in which the Denises were working, we might also look to other television programmes made in Kenya during the same period to contextualise the style of address of the Denises' wildlife documentaries. For example, there were a number of East African adventure series made for independent television in late 1950s, by companies like Grose Krasne (*Jungle Boy*, 1957; *African Patrol*, 1958–59) and Beaconsfield Productions (*White Hunter*, 1957–58). Whilst the Grose Krasne series was shot entirely on location in Africa, *White Hunter* was shot mainly in the studio in Twickenham, with some (mostly stock) location footage to establish the Kenyan setting. Whilst it is difficult to assess the impact that these programmes might have had on British views of Kenya in the late colonial period without being able to view the series, or the extent to which they drew on the Kenyan imagery established in the Denises' earlier programmes, episode synopses suggest an emphasis on danger and spectacular action for their mainly white protagonists, who were adept at negotiating a colonial world populated by stereotypically 'good' and 'bad' Africans. *Jungle Boy*, shown in a pre-teatime slot in 1957 on a Sunday, and repeated in 1960 (at 17.15 on Saturdays) and again in 1969 (at 16.55 on Mondays), falls very much into the category of the white boys' adventure series discussed by Jeffrey Richards (1986), which were popularised in the late nineteenth century and described by Barish Ali as conjuring up 'an incredible promise of fame, riches, and escape' (2011: 1141). Similarly, discussing popular culture of the 1910s Jan Nederveen Pieterse concurs: 'Africa came more and more to resemble a vast recreational area, an ideal setting for boys' adventures. Fiction for entertainment and for the young follows the colonial matrix' (1992: 111). Like the Denises' programmes, this depiction of the Kenyan jungle in which people and animals were shown cooperating (against threatening black *and* white enemies in the case of *Jungle Boy*, at least) created a narrative of simultaneous excitement and racial harmony (and harmony with Kenya as a place, as represented by the space of the jungle and its fauna) at a crucial stage in the history of British colonisation of the country. Given that *Jungle Boy* was broadcast in a family viewing slot (as was *African Patrol*, with all but the most risqué episodes of this series being broadcast at 18.10 on Wednesdays), the appeal of these spectacular 'colonial adventures' as family entertainment might be seen as an attempt to *normalise* the white family's position in East Africa in this period. The programmes of Armand and Michaela Denis function in the same way.

Armand and Michaela Denis: At home on safari

Armand and Michaela Denis produced and presented a range of popular wildlife programmes for British television between 1954 and 1965: *Filming Wild Animals* and *On Safari* for the BBC and *Michaela and Armand Denis* and *Armand and Michaela Denis* for ITV.[16] Their double act was characterised by her breathy, very English, enthusiasm, and his image as a knowledgeable, European, older man.[17] In their interaction with each other, Michaela frequently took on the position of a diegetic spectator and, in doing so, mediated between the 'wild' outside world of Africa and the viewers at home. In an article introducing their first one-off broadcast, *Filming in Africa*, for the BBC, Head of Television, Cecil Madden, sums up the appeal of the Denises:

> In these hard times of travel and currency restrictions and unpredictable weather, television's window on the world can bring the remote places of the earth to your living room on a magic carpet by armchair travel [...] The Denises' cameras open up the living world of the animal kingdom [...] in a new and personal way.
>
> (Madden, 1954: 7)

This idea, that the Denises provided the viewers with a vicarious experience of a dangerous postwar world (subject to 'travel restrictions and unpredictable weather'), implicitly giving access to 'the colonies' to viewers from the comfort of their living rooms, was the key to the presentational style of their programme making.[18] Their programmes epitomised the spectacular space-binding qualities of television which broadcasters were keen to promote in the 1950s: the utopian promise of being at home and away at the same time – as Lynn Spigel has argued in relation to US television of the same period (1997: 213). The fact that the Denises' programmes offered to take ordinary viewers at home 'on safari' might therefore be seen as public service: this was a space that was financially and geographically out of reach for most, as discussed above.

The moments of the Denises' first series for the BBC, *Filming Wild Animals*, that frame the broadcast of their location footage take place in a domesticated studio space (which borrowed set dressing from the cosy afternoon chat show for women, *Joan Gilbert at Home*, BBC, 1952–4).[19] Dressed in evening wear, and sitting side by side on armchairs as if addressing guests, in front of a 'window' with smart venetian blinds, the Denises' space-binding direct address to 'you the viewer at home' configures their programme as a kind of interstice in which the 'outside world' of the 'animal kingdom' and colonial Africa, and the 'interior world' of the domestic viewer in Britain are brought together. At the close of the programme, when describing how much they have enjoyed being in London, Armand expresses his yearning to return to Africa in the following terms: 'Now, when we travel to Africa, we feel that we're not alone. You are all travelling with us through the country that we feel is the most beautiful and most exciting.' Television, in this light, might be seen as the ultimate colonial apparatus, as a medium which enables one to be at home and abroad at the same time. What I am referring to here is not, of course, the literal process of colonisation in relation to the history of empire (despite the historical and geographical context in which the Denises were working), but rather the idea of colonisation as a metaphor which has been frequently employed by television scholars in order to understand the way that television enters the home. Arguably television has always been seen as a medium of colonisation; many of the meta-narratives of television studies rest on the notion of television as *colonising* domestic space. What Milly Buonanno calls the '*doyenne* of televisual metaphors' (2008: 17), the idea of television as 'a window on the world', might therefore be seen as a colonial idea:

> The ambivalence of the relationship between internal and external, inside and outside, that characterizes [...] the process of introducing the television set into private living spaces, is here

reproduced in the phenomenology of a symbolic and imaginary departure, favoured by the same medium that so to speak 'colonized' these spaces.

(ibid.)

Television studies is thus shot through with the discourse of metaphorical colonialism, whether it be via the window-on-the-world metaphor,[20] or through the powerful idea of 'witness' via television, or, indeed, in the notion of co-presence which informed so much of our understanding of the specificity of television viewing and address in the pre-digital age.[21] Given the contemporaneous struggle for independence in Kenya, there is a huge political significance in the 'at home abroad' presentation that characterised the Denises' wildlife programmes. Jan Nederveen Pieterse has argued that 'above all, the safari is a demonstration of European mastery, of the superiority of western technology, and a crucial symbolic episode in the colonisation of Africa – the "dark continent" is manageable' (1992: 112).

One of the things which is so striking about these programmes is that the spectacular 'looking relations' within them are so expressive of the relationship between the British and the Kenyan people at this pivotal moment in history. Even when we are being shown landscapes and wildlife, rather than townscapes and people, we are aware of the power relations at play in the act of looking and the act of filming, of seeing and of 'capturing', and of the Denises' (perhaps subconscious, perhaps overt) desire to reinforce these relations as natural or inevitable. Depicting a land which is 'wide open' to them, the Denises' programmes perhaps show that the myth of Kenya as a 'new Australia' persists. E. Ann Kaplan has argued, leading on from bell hooks' suggestion that 'there is power in looking' (hooks, 1999: 307), that 'subjects in a culture are […] constituted as able to "see" or not' (Kaplan, 1997: xvi). Unpicking the imperial gaze, Kaplan is particularly interested in the ways in which 'looking relations are determined by history, tradition, power hierarchies, politics and economics' (ibid.: 7) and argues that 'travel implicitly involves looking at, and looking relations with, peoples different from oneself […] [it] provokes conscious attention to gender and racial difference' (ibid.: 5). In the Denises' programmes, we see a visualisation of the looking relations described here by Kaplan.

In the episode of *Filming Wild Animals* discussed above, for example, a sequence in which a young elephant is captured encapsulates the issues of who sees, who speaks and who is seen in colonial television. Following the opening address to the viewers in the studio, Armand turns directly to camera and begins to describe how they captured the animal. The opening shot of the sequence that follows is a medium long shot of the Denises sharing a pair of binoculars and looking for the elephant from their vantage point on a dead tree. Following this, on Armand's statement in voiceover that 'On just about the third day we were there, I happened to be looking over the country and I saw this little thing', a cut is made to an iris shot (Fig. 3.1) of the elephant from Armand's point of view (which is confirmed by a reverse shot of him looking through binoculars again). In cutting from studio to filmed footage (from Armand's face to location shot), the sequence is established as being taken from his subject position, if not entirely from his optical point of view, and his voiceover confirms this. This reaffirms that this rendering of the landscape as spectacle is his (white, European, colonial) perspective of Africa, a perspective the viewers at home are invited to share. Furthermore, the Denises are simultaneously *owners of* and *placed within* the imperial gaze in this sequence. Throughout this, and their later series, they are placed within the landscape, and constantly seen on camera (often operating a camera themselves or looking through binoculars), reinforcing the notion that this is their 'rightful/natural place'. By rendering this land into landscape, they take symbolic control of the Kenyan countryside. At the beginning of the elephant capture sequence, the iris shot uses an early cinematic device to make a point about looking relations and the Denises' mastery of the African landscape. This iris framing draws attention to the *constructedness* and the spectacular nature of the image: 'Africa' therefore becomes a colonial construct in these safari adventure narratives for television. This is then re-emphasised throughout the sequence by shots of Armand pointing during the capture of the elephant; the camera either follows his direction, as do the other 'characters', or we cut to a point of view shot, which reaffirms him as the bearer of the imperial gaze.

The story of the sequence – the capture of the elephant – is a story of collaboration between Europeans and Africans, and here this sequence can be read as a metonym for the Denises' programme-making as a whole, as well as a metaphorical image of colonialism which, in the years of struggle

At home on safari

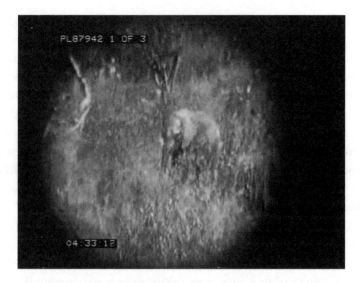

Fig. 3.1 The colonial view: *Filming Wild Animals* (BBC, 1954)

for independence, white settlers might be keen to promote. As elsewhere in the series, black Africans are depicted here as voice-less workers who move as a mass: they are a group, not individuated (unlike the white men and woman); they are not named, and their presence is often not acknowledged by the voiceover. For example, in Armand's voiceover about the capture of the elephant ('And the first to grab it of course was David Sheldrake himself, but very soon I came to his help, and then Michaela'), there is no mention of the other men on screen restraining the elephant: they are invisible to Denis, if not to the viewers.

Michaela's position in the sequence is interesting. As in the 'pets at home' sections of their programmes, where she is shown introducing 'tame' wild animals to the viewers and which are striking in the extent of the anthropomorphism at play, Michaela is depicted as a 'civilising' and 'domesticating' figure through the taming of the elephant. Furthermore Michaela is framed in this sequence, at the front of the shot, surrounded by Kenyan men, or running amongst them, in a way that emphasises her whiteness and her femininity. Her whiteness, in particular, gives her visual pre-eminence in the frame in this sequence, her pale skin giving off what Dyer has referred to as the 'glow of the white woman' (1997). In this sequence, as elsewhere in their programmes, Michaela's gender and race render her a troubling figure, which is perhaps explained by Sara Mills' suggestion that the female travel writer challenges the 'rules concerning [British middle-class women's] movement in the public sphere' (2005: 3); this idea might be transposed to think about the female programme maker who simultaneously confirms the norms and values of European femininity whilst challenging them through her activity, her sense of narrative agency and her position *and movement* within the 'wild space' of colonial Africa. We might also understand Michaela's sense of agency as signifying a certain kind of modernity in this moment; this is certainly how Wendy Webster reads analogous fictional characters in the postwar film about Africa (2011).[22]

In fact the Denises both conform to colonial Kenyan stereotypes outlined in Richard West's book, *The White Tribes of Africa* (1965). Their constant emphasis on their experience and ease in the wild areas of Kenya figures them as 'white hunter' types, urbane enough to converse with other white Europeans but also skilled at 'dealing with the natives' and the native flora and fauna. This is a generalised African colonial type found in the literature of Kipling and Haggard: 'a self-reliant, humorous, compassionate, brave and versatile character [that] emerges by the end of the [nineteenth] century' (Lewis and Foy, 1971: 28). Furthermore, Michaela is figured as alluring and sexualised (she was famous for her red hair and her silver nail polish, and for the fact that she looked good in her figure-hugging khaki shorts), a contradictory figure: domesticating and at the same time sexually alluring/dangerous; cultured and urbane and yet 'wild'. These contradictions were frequently echoed by the ways in which she was treated in the press,[23] but were also present in the Denises' programmes. A moment which encapsulates this dichotomy can be found in the 1958 *On Safari* episode 'Search for Gertie', in which Armand and Michaela are shown at home in the living room of their bungalow (typical of the 1950s African colonial style). The episode, shot in colour but shown in black and white, contrasts the Denises' familiar and urbane furnishings (Wilton rug, chintz curtains) and Michaela's blue tea dress, with the sight of their serval (a small wild cat) pouncing at Michaela's feet whilst Armand announces that 'I have given up trying to make a house cat out of serval but Michaela hasn't. She believes that what wildcats need is love and that with enough love, everything will be fine.'

Throughout both series, the 'domestication' of wild animals is spoken about in terms that echo colonial discourses of racial harmony or the 'domestication' of Africa, particularly in relation to Michaela's knack for taming animals (her first two biographies were titled *A Leopard in My Lap* (1956) and *Ride a Rhino* (1960)). In the elephant capture discussed above, emphasis is placed on the *acquiescence* and the *consent* of the animal to its capture: 'And the little one marched into his enclosure as if he'd been born to this. And we knew he'd be tame and he'd be alright'. Perhaps it is a rather obvious point to make that the anthropomorphising of animals in these programmes frequently echoed a rather elegiac reflection on the relationship between the coloniser and the colonised. Throughout the Denises' programmes, their 'pets at home' sequences figured wild African animals (baboons, chimpanzees, a serval, a mongoose, etc.) as 'funny little children', living in harmony with their captors and each other. In an episode of *On Safari* broadcast on the 21 May 1958, Armand states that

> On the whole, it's remarkable [...] how many species manage to live peaceably with each other and how many species can be tamed sufficiently to be a pleasure to have around [...] To keep wild pets means taking on a certain responsibility to see that they have conditions of living to which they can adapt themselves.

It is easy to hear echoes of paternalistic colonial discourse in this dialogue (notions of patronage, responsibility and a 'civilising' drive).

The Denises did not, however, confine their programmes to the exploration of the 'animal kingdom'. Many episodes also had a distinctly anthropological focus. When indigenous people were shown in the Denises' programmes, they were depicted either as being odd or alien in some way, or in need of education, particularly about conservation issues, as subservient to white settlers or, as Wendy Webster puts it in relation to the depiction of Africans in British cinema of the same period, as 'part of the wildlife' (2011: 239). This confirms Fatimah Tobing Rony's critique of ethnographic spectacle as a form of spectacle 'in which indigenous peoples are exhibited and dissected – both visually and literally' (1996: 190). The lack of diegetic sound on most footage in the Denises' programmes renders black Africans without a voice; they become literally a narrated image, the object of the imperial gaze, without agency and subject to what Aimé Césaire calls the 'thingification' of colonial discourse (2000: 42); in the natural history programme of the 1950s, the subaltern literally cannot speak. In the final sequence of the episode of *Filming Wild Animals* discussed above, in which the Kamba tribe (referred to as the Wakamba in this programme) are shown weaving and dancing, montage editing is used to emphasise a collectivity which denies the black African a sense of individuation. The voiceover at this point in the episode stresses the Kamba people's *pleasure* in the presence of their white colonisers: 'the arrival of white people in the village is always the occasion for a lot of excitement and a lot of rejoicing and almost invariably a dance begins'. There is no return of the scrutinising gaze of the Denises' camera here and the presence on screen of Denis and his crew during the dance sequence underscores their elevated position of power as authors and controllers of the look. Kaplan has argued that 'since blacks are not constituted as subjects, they cannot look (i.e. look *for* whites, satisfy openly their curiosity about whites) let alone gaze (in the sense of dominating, objectifying)' (1997: 7). Therefore, when facial close-ups occur in this sequence, as in a moment where Michaela speaks *for* a woman weaving a bag, the composition of the image sees the weaving woman look beyond and away from the camera, not meeting, *and not challenging*, its invasive and scrutinising gaze. It is also telling that the native Kenyans who appear in the Denises' programmes in this period in colonial history are mainly confined to the Kamba and Maasai tribes. Unlike the more urbanised, educated Kikuyu people, who were almost wholly responsible for the Mau Mau uprising (according to most reports), the Kamba were seen as a 'martial race' and ready to collaborate with white colonials.[24] On the other hand, as Richard West claims, the Maasai were romanticised as 'noble savages' (1965: 36–7). It is easy to see how these romantic stereotypes were drawn out for the television-viewing public by the Denises' programmes for a large portion of the 1950s, and how representation of Kenyan people was reserved for tribes who were not radicalised or wholly involved in the struggle for independence.

More often, however, African landscape was shown as empty wilderness, 'protected' or rather 'guarded' by the colonial administration. Here the landscape becomes a symbolic spectacular geography of owned space in which colonial ownership is emphasised by the expansive movement of the Denises' cameras: the mobile camera of the safari documentary thus repeatedly and symbolically colonises space (to return to a metaphorical understanding of this term). Particularly in the later series, *On Safari*, the Nairobi National Park becomes a symbol of simultaneously wild, 'empty', but colonised space. For example, in episode one of this series, Armand exclaims that 'African National Parks are like no park that you may be familiar with. They are huge areas, as large as English counties, the wildest country left in Africa, set aside for the protection of wildlife' and he bemoans the fact that soon 'There may well be nothing left of the old, fabulous Africa for our grandchildren'. This latter statement is accompanied by a slow, panoramic track around a lake, with Denis explaining that 'this is the view from the veranda of our cabin' (see Fig. 3.2). Here it is clear that this is a *colonial* spectacle, to be protected for the viewing pleasure of white Europeans rather than for the good of the African people; however, there is contradiction within this episode on this issue. When showing the park director addressing a group of Kenyan schoolchildren later in the episode, synchronous sound allows his speech to be heard: 'This is your park, these are your animals, and we are preserving them for the future of your own people as well as for the whole of mankind. This is a most important part of your heritage'. Again, Armand counters this by saying that Africans have never thought of animals except in terms of 'so much meat' and Michaela confirms this by saying that 'To the uneducated African, the thought of protection or conservation of animal life is incomprehensible. Why keep all this meat instead of eating it all at once?' It is clear then that the highly problematic broader address of this series supposes that the landscape and wildlife of Africa are more important in their value as (televisual) spectacle for the 'home viewer' than they are to the Kenyan people. As Beinhart and Coates state of the Denises' programmes (and those of George and Marjorie Michael[25]), 'So convinced were these conservationists of the righteousness of their cause that they, reflecting broader colonial preoccupations, overlooked claims that indigenous peoples might have on reserved land' (1995: 84).

Perhaps these programmes are a rather easy target for critique, given the context of their production, and their now wholly unacceptable treatment of Kenya and its people. What this chapter opens up, however, is

Fig. 3.2 The colonial view: *On Safari* (BBC, 1957)

the previously underexplored idea of colonial television and the ways in which programme-makers within the colonies figured Kenya, in particular, as spectacle for the viewer 'at home': in short, television contributed to the imagining of Kenya for the audience at home in the UK, and resonated with the characterisation of the country, its people and its landscape during the late colonial period elsewhere within popular culture. Analyses of editing, camerawork, sound design and narration show that natural history television is never a politically neutral genre, but rather that in rendering landscape, wildlife *and people* as spectacle it eloquently reveals a great deal about global cultures.[26] Ultimately, what is also illuminating is that the Denises failed to adapt to the changing landscape of British television and the increased 'sophistication' in wildlife broadcasting. Correspondence in Armand Denis' file in the BBC Written Archives shows that from the beginning of 1962 onwards (i.e. just short of two years before Kenya's independence from Britain), Denis was being asked to drop the 'pets at home' sections of their programmes, something which he appears to have been reluctant to do as this was something of a signature style. In February 1964, Nicholas Crocker at the Natural History Unit writes in exasperation to Denis that:

the present day television audience will not readily accept this sort of pets treatment any longer. If they are going to accept it, you have really got to dress it up very carefully [...] People would like to know not just that you are keeping them as pets, but that you are studying them most carefully. The television audience does not take too readily now-a-days to an anthropomorphic approach i.e. pets' names etc. They want to know about animals as animals, but not so much about animals as extensions of human activity.[27]

Given that, as it has been explained in this chapter, this 'at home on safari' address and the diffusion of colonial discourse into discussions of animal behaviour were key to understanding the Denises' programme-making within the late colonial period, implicit in Crocker's criticism, particularly in the final sentence of this extract, is that post-independence they were seen as representing outmoded views of Kenya. Television had entered a new age in its story as a colonial apparatus, and in its depiction of the spectacle and beauty of the natural world, and left the Denises behind.

4

Visual pleasure, natural history television and televisual beauty

This chapter revisits my earliest work on natural history television, public service broadcasting and visual pleasure as a springboard for a consideration of the idea of beauty as it relates to television. It interrogates this concept, and considers how beauty has been identified as one of television's key pleasures, exploring the ways in which beauty (beautiful programmes, beautiful moments in programmes) has been defined in television policy and production, as well as in the discourses surrounding television programming (in reviews and in viewer evaluation of programming). Beauty is a (largely unexplored) criterion in the assessment of 'quality' television; this chapter seeks to unpack its use within and inside of the television industry in order to better understand how and why the beautiful is valued in relation to TV.

Throughout the history of British television, regulators, critics and scholars have struggled to find a single definition of what quality public service broadcasting is. Historically, the question of which programmes might epitomise quality and the public service broadcasting ethos has been afforded different levels of importance at various points throughout this debate. For example, reflecting on his role on the Pilkington Committee into the state of British broadcasting in the early 1960s, Richard Hoggart stated that:

> The Pilkington Report has been criticised for not naming the programmes it admired. I understand it isn't the thing to do in a Government report. Personally – probably other members felt the same – I would have liked to use many particular examples.
>
> (Hoggart, 1962: 9)

On the other hand, three decades later, following the publication of the 1988 White Paper *Broadcasting in the '90s: Competition, Choice and Quality*, Charlotte Brunsdon identified a very different trend in the industrial and regulatory rhetoric of 'quality' when she noted that

> One of the ways of side-stepping the difficult critical arguments about quality television, whilst still being able to attest to its existence and virtue, is to refer to specific programmes, which can then function as shorthand for taken-for-granted understandings of 'quality'.
>
> (1997: 139)

The article on which this chapter was initially based (Wheatley, 2004) offers an analysis of the BBC's *The Blue Planet* (2001) and the rhetoric surrounding it. It seeks to unpick some of those 'difficult critical arguments' about quality television in relation to a specific example from the natural history genre, by highlighting the underacknowledged role that questions of visual (and aural) pleasure played in debates about public service broadcasting in Britain at the beginning of the 2000s. I proposed that those people repeatedly using *The Blue Planet* throughout 2001 and 2002 to argue for the continued vigour of public service broadcasting in British television – including Greg Dyke, then Director General of the BBC; Tessa Jowell, then Culture Secretary; and Patricia Hodgson, then Head of the Independent Television Commission – inadvertently announced the state of disrepair of the rest of British public service broadcasting. I therefore argued that theories of flow needed to be reintegrated into the debate around quality television and that moments of heightened visual and aural pleasure on television serve only to highlight the relative aesthetic impoverishment of the rest of television's broadcast flow. This chapter assembles extracts of that article, leading into an extended consideration of beauty as a criterion of evaluation on television, in which I discuss the meanings of this term in relation to television more broadly.

The Blue Planet, an eight-part BBC/Discovery co-production about the world's oceans, was broadcast in the autumn of 2001 and was, by many of the standards set for television, seen as a great success. Quantitative measures of success include the fact that some episodes garnered just over 12 million viewers, it had the BBC's fourth-largest viewing figures for 2001 and was sold to over 50 countries worldwide. The programme was also seen as a critical success and was well received in the British press. It was followed, in September 2002, by the BBC's *The Blue Planet Prom in the Park*, an open-air screening of extracts from *The Blue Planet*, which formed part of the Proms (a series of classical music concerts and recitals traditionally held throughout the summer), accompanied by a live orchestra and choir in Hyde Park, London. At this event, I began some research into the viewers and viewing practices of this genre by conducting 65 brief interviews in the park (interviewees selected at random whilst queuing to go into the event), followed up by a much longer set of questions in questionnaire form, returned by 28 of the people I had interviewed. This research informs and illuminates the following discussion of the role of visual pleasure in the quality television debate in that many of those responding in the interview and questionnaire articulated their experiences of viewing the natural history genre on television from within the quality television/public service broadcasting debate. These particular viewers (who had, admittedly, already singled themselves out as 'discerning viewers' by attending a live screening of series highlights accompanied by the BBC Concert Orchestra and the choir of Magdalen College, Oxford) were well versed in the rhetoric of the debate and many felt able to speak with confidence about the visual and aural markers of 'quality' within this series. The following analysis therefore seeks to establish how and why *The Blue Planet* came to be the most oft-cited example of quality public service television in the early part of this decade, when notions of public service broadcasting were being brought into question by industry regulators, media professionals and television viewers alike. As John Street has argued – against Tony Bennett's depiction of aesthetic discourse as 'a really useless form of knowledge' (1985: 34) – the question of aesthetics must be paramount in debates on quality television within a public service system, given that 'the policy decisions which result in subsidy for works of culture or quotas for broadcasting are also aesthetic judgements about the value of particular cultural forms' (Street, 2000: 28).

[...]

Inserting natural history into the quality debate

The flurry of critical activity around public service broadcasting and 'quality television' in Britain at the beginning of the 2000s centred, for the most part, on the transformation of the BBC within the context of a multichannel, interactive, pay TV service. The apparent disparity between the public service ethos of British television to that point and the BBC's empire building and 'rampant commercialism' was being constantly raised in the trade press and can be summarised by New Labour peer Lord Lipsey's comment that the remit of the BBC under Greg Dyke was to 'colonise, compete and destroy' (Keighron, 2002: 13) or the inflammatory suggestion by the then ITV Director of Programmes David Liddiment that Dyke had no grasp of the BBC's public service responsibilities (Maggie Brown, 2002: 2).

At the intersection of industrial and critical discourses, debates around quality television have attempted to reinscribe (or, rather, illuminate) questions of aesthetics within industrial and regulatory debates. The critical activity that surrounded the 1988 White Paper and the call for its ill-defined 'threshold of quality' to be clarified are extremely informative in the context of a discussion of the changing face of natural history, particularly given that this is one of the few areas of Television Studies where notions of visual and aural pleasure have been discussed. To give a prominent example, Brunsdon's essay 'Problems with Quality' ([1990] 1997) unpicks the application of the quality label to the heritage drama (specifically, *Brideshead Revisited*, Granada, 1981; and *Jewel in the Crown*, Granada, 1984) by isolating four essential components of 'quality': money, heritage export, the best of British acting and a literary source. These 'components of quality', at first glance specific to the heritage genre, can in fact also legitimately and usefully be applied to an analysis of *The Blue Planet*, with the 'literary source' in this case being replaced by 'proven scientific fact' and the presence of expert discourse in the series, just as the best of British acting might be substituted, in the form of David Attenborough, by the best of British presenting. Of course, Attenborough is a highly significant figure within both the natural history genre and the BBC as a whole and, it can be argued, embodies in his celebrity persona (reserve coupled with a 'typically British' sense of eccentricity) and style of delivery/performance (particularly the whispery reverent timbre of his voice) a number of the associated 'virtues' of quality public service television.

To dwell a little longer on the question of money, Brunsdon writes:

> Like MGM musicals, both these series [*Brideshead Revisited* and *Jewel in the Crown*] cost a lot, and, as importantly, looked as if they cost a lot […] [But] throwing money at a project doesn't guarantee that it will look expensive, and it is the combination of restraint and uncommon spectacle which is the key signifier here. These were expensive productions in which the money was spent according to upper-middle-class taste codes.
> (1997: 142–3)

If quality television, according to Brunsdon's analysis of the discourse of quality, is understood to be that programming which is conspicuously expensive according to 'upper-middle-class taste codes', then *The Blue Planet* fits this description perfectly. Taking the opening of the first episode in the series as an example, we can isolate a number of indicators of this coded expense within the sound and image tracks. As the title sequence runs, showing a succession of iconic shots from the series and establishing the muted colour palette of blues and greens that pervades *The Blue Planet*, the score acts as the most immediately striking indicator of expenditure, according to upper-middle-class taste codes, not only because the composer George Fenton is associated with the British cinematic epic (e.g. *Gandhi*, Richard Attenborough, 1982; *Cry Freedom*, Richard Attenborough, 1987), but, more immediately, because the music has been scored for and performed by a full orchestra, with all the pomp and ceremony that this entails. The use of the choir from Magdalen College, Oxford, is also telling. Music in this opening sequence therefore quickly connotes expense according to upper-middle-class taste codes and a budget that allowed for the employ of a feature film composer and a well-respected choir from a long-established British institution. However, if we also look at the close matching of sound and image (such as the lengthy build up to the crescendo that accompanies the emergence of the tail of a blue whale in the first sequence of the episode proper), it is also strikingly evident that this is a score *specially composed* for this series. Even if we were unaware of the calibre of the composer and choir (a difficult feat, given the stress placed on their involvement in interviews in the press and listings guides, interviews with the producer and composer in the DVD extra footage, and so on), the close match between image and sound acts as a further subtle indicator of quality. As the whale's tail heaves slowly out of the water, the score builds towards a fanfare of horns which trumpets the tasteful expenditure of the programme.

Furthermore, the filming and editing in this opening sequence support this sense of awe with dignity, or rather expense with uncommon restraint. The pace of the editing at this moment is languorous, as the blue whale is repositioned in a number of extreme angles and shot both extremely close up and extremely far away from a bird's eye view, with each shot in the sequence emphasising the sense of greatness, size and depth. It is significant that *The Blue Planet* does not begin with chase scenes or computer generated images (CGI), as did its natural history contemporary *Walking with Beasts* (BBC1/Discovery/Pro 7/TV Asahi, 2001). Without these televisual 'whistles and bells', the shots of the ocean in the opening sequence continually emphasise depth and breadth as well as a much slower pace, again signifying by comparison the series' sense of restraint or reserve. As such, the blue whale in this opening sequence can be seen as a metonym for the entire series, a representative of *The Blue Planet*'s expansiveness and reserve, with the temporally long shots emphasising its slow movement, suggesting that the series will 'take its time too'.

[…]

Visual pleasure and public service broadcasting

When questioning the kind of public service that the makers of *The Blue Planet* were seeking to provide, it is instructive to look at the way that the series was presented in the BBC publication, the *Radio Times*, where it received extensive coverage throughout the duration of the series, often featuring in viewers' letters as well as in-feature interviews, competitions, advertising and the 'pick of the day' sections of the magazine. In both editorials and, frequently, in the letters pages, *The Blue Planet* was praised and celebrated for being spectacular, sublime, choreographed, balletic – even cinematic – with the *Radio Times* promoting the belief that the spectacular nature of the series, the visual and aural pleasure provided by the programme, was a public service in itself. For example, the magazine's editor Nigel Horne described his first reactions to *The Blue Planet*:

> Recently, I was lucky enough to attend a sneak preview of BBC1's new documentary series *The Blue Planet*. As the credits rolled and we applauded loudly its sublime photography and dramatic narration, it was hard not to think: 'At last, television doing what it does best.'
>
> (2001: 7)

In a similar vein, one week later, viewer Judith Proctor wrote to the *Radio Times*:

> *The Blue Planet* is quite simply the most stunning piece of television I have ever seen. Choreographed like a ballet, with perfect music, showing the oceans like I've never imagined them before. This is worthy of a cinema release – it demands to be seen on the big screen.
>
> (2001: 6)

This gushing enthusiasm from both editor and viewer, using the terminology of the *cinematically spectacular* to praise the series, might be seen as rather confused or confusing; on the one hand *The Blue Planet* is seen as the 'best of television' and, on the other, it is seen as cinematic (or, rather, 'not television') in its greatness. [...] It is evident that the praise for the series cited above comes from a critical commonplace in the reception of and understanding of television as being an unspectacular medium; here, I refer to those well-versed arguments about television being sound-led or being a medium of communication rather than a site of visual and aural pleasure. From the beginnings of television studies – certainly since John Ellis' *Visible Fictions* (1982) – the dichotomies of information and spectacle, television and film, have been emphatically stated and restated. In fact, changes in television viewing call for these dichotomies to be radically reassessed and the dramatic increase in television screen size, sound quality, picture clarity and so on may make the extreme visual and aural pleasure of *The Blue Planet* 'worthwhile' and medium specific in the context of today's viewing practices. In essence, *The Blue Planet* is perfectly designed for this 'Bang and Olufson' viewing culture. In the programme for *The Blue Planet Prom in the Park*, composer George Fenton writes that 'the boundaries between cinema and television are becoming progressively blurred [...] [M]ore and more people are able to enjoy television on high quality screens and with stereo sound to rival cinema'. Fenton's statement suggests that *The Blue Planet* was made for broadcast in 2001 with precisely this technologically enhanced form of television viewing in mind.[1] Of course, this ultimately raises issues of accessibility to the pleasures of *The Blue Planet*. Just as James Bennett (2003; 2008) argues that many of the pedagogical elements of early interactive natural history programming were problematically unavailable to all (due to their being embedded in costly digital delivery platforms) [...], so the audio-visual pleasures of *The Blue Planet* are best enjoyed by those with the capital (both cultural and actual) to appreciate them.

It is significant that *The Blue Planet* was a collaboration between the BBC and the Discovery Channel, although an examination of the branding of the programme in the UK finds that the presence of the latter as co-producer is suppressed to a great extent. [...] This co-financing means that it was perhaps disingenuous of Greg Dyke and Tessa Jowell and others to claim *Blue Planet* solely as a licence fee success; the programme was only part-funded by the licence fee, with money also being provided by an American TV company. International co-production or co-funding is not a new phenomenon in the natural history genre, particularly for the Attenborough-led mega-series. The first of these series, *Life on Earth* (1979) [...] was co-produced by the Natural History Unit, Warner Brothers and the German Reiner Moritz Productions, for example, and was sold to more than 100 countries and seen by an estimated 500 million people worldwide. John Caughie's work on the heritage drama and international co-production is illuminating in relation to the balancing of commercial and public service imperatives and the move from 'the necessity of the national to the demands of the international' (Caughie, 2000: 206). As he notes, international co-production allows a 'scale of ambition in terms of production values which few public service broadcasters [can] emulate' (ibid.: 208), which leads us to question the impact of this economic context on the aesthetics of the genre, to ask whether this spectacularisation of British natural history is embedded in its co-funding structures and to consider whether the 'requirements of the international market [are] compatible with the interests of the national public which public service broadcasting is there to serve?' (Caughie et al., 1986: 103). [...]

In [...] *The Blue Planet*, there seemed to be a much greater emphasis on abstract visual and aural pleasure when watching the series than there had been in earlier examples of this genre, on moments that were isolated from the narrative thrust of each episode or the series as a whole. It was also these moments – where the *beauty* of what was being shown was emphasised above all else and where Attenborough's voiceover was lost in favour of a soaring orchestral score – that were marked out as moments of 'quality' in the discourse surrounding the programme and were constantly emphasised and re-emphasised by a selection of short clips that represented the programme on others shows, in magazine feature illustrations and in advertising spots on the BBC (such as underwater footage of a 'baitball' of sardines being attacked by marlin, dolphins, sharks and a Sei whale, or a deep-sea montage of a team of alien-looking underwater creatures ascending to feed). [...] It appeared to me that there was an even greater emphasis on abstract visual and aural pleasure outside of an educational remit in *The Blue Planet* than in the programme that initiated this cycle of Attenborough-led mega-series, *Life on Earth*. To test this hypothesis, I conducted a quantitative analysis of several episodes from both series and measured the ratio of narrated and non-narrated time, discovering that, on average, *Life on Earth* was 57 per cent narrated whereas *The Blue Planet* was 41 per cent narrated. Aside from these statistics, when Attenborough's voiceover is present in *The Blue Planet*, it acts in much the same way as George Fenton's score; it is 'pasted onto' the series at the end of production to act as an aural indicator of quality and distinction [...]. Whereas Attenborough was present in other of these Natural History Unit mega-series, speaking directly to camera from the edge of a swamp or halfway up a tree, in *The Blue Planet* he largely remains within the confines of the recording studio.[2]

To give an example of a moment from the series that represents this greater emphasis on visual/aural pleasure outside of the programme's didactic remit, in the fifth episode, 'Seasonal Seas', a lengthy montage of several kinds of jellyfish is particularly striking. In this sequence, edits do not necessarily serve to give us a better view of one particular species or behaviour in the jellyfish, but rather move from one visually stunning shot of this marine life to the other. There is little logic of exposition here and many of the edits are non-instrumental, moving from one abstract image to the next. Once Attenborough's voiceover begins the sequence by announcing, 'Jellyfish. They may appear to be delicate as well as beautiful, but they are deadly hunters', we see seven separate shots of five different species without any further exposition of behaviour or type. As the sequence continues, minimal narration about the feeding habits of jellyfish, which is not always closely tied to the image, is offered, again followed by an extended montage of this ethereal marine life accompanied only by sound effects and music. Here the score suggests [...] that these are strange and unknowable creatures by referencing the diegetic and extra-diegetic sounds of science-fiction film and television, thus subtly contradicting the didactic aims of the series [...]. This jellyfish sequence, and others like it, can be read as a moment of non-narrative visual and aural pleasure that exists outside of the informational remit of the programme and that would seem to confirm that the 'service' being performed by *The Blue Planet* is not just one of education, but rather one of entertainment [...], and possibly relaxation, the provision of space and time in which the viewer might figuratively 'switch off'. In fact, whereas it has been argued that this series has appeals that are cinematic rather than televisual (in the traditional senses of both words), this sequence, and others like it, might actually be seen as aesthetically closer to commercial relaxation videos, with their repetitive waves of tropical fish or 'cosy' log fires playing on a loop.[3]

The series' abstract visual and aural pleasure was continually emphasised in the reception of the programme, with the audience research I conducted seeming to confirm David Morrison's suggestion that 'most viewers are sophisticated genre readers and have their own version of the "quality" judgement, dividing programmes into the worthy and the trashy, and taking pleasure from both' (Frith, 2000: 46). The viewers I talked to and received questionnaires from seemed to be acutely aware of both changes in the genre and the kinds of pleasure that they derived from the series itself, as well as being well versed in the rhetoric of 'quality' television. To give an indicative example, a 27-year-old man from Glasgow, commenting on recent changes in natural history programming, offered the following:

> I think there has been a definite move towards more sensational natural history, less in depth. 'Planet's most deadly maneaters' and whatnot. I don't hold this to be an entirely bad thing if the 'tabloid' natural history programmes can recruit and entertain a broader audience and as long as there remains a place for the slightly more thorough work (exemplified by the BBC) where I feel there is more wonder and reward in the long term. But then I like symphonies and test cricket, so I am clearly the sort of person who enjoys life's deeper, less immediate pleasures.

What is evident from this statement is that this particular viewer both sees himself as a discerning connoisseur of middle- to upper-middle-class culture (indicated here by test cricket, symphonies and the BBC) and is evidently articulate in debates about natural history television, public service broadcasting and visual pleasure. This comment was not given in isolation, with other viewers commenting on the score ('breathtaking music, I could listen to it on its own', 55-year-old teacher from Suffolk) and spectacular footage ('very good filming. That's why they win so many awards. Great camera work', 31-year-old secretary from London), again offering a clear sense of judgement from within the television industry's discourses of quality (specifically, in the above cases, the presentation of ubiquitous television awards and the possibility of sell-through CD merchandising).

–

Beautiful television

The public service of the provision of relaxation and visual pleasure offered by *The Blue Planet* was particularly emphasised by the fact that the broadcast of the first episode of the series came on 12 September 2001, the day after the terrorist attacks on the Twin Towers in New York. Writing in the letters page of the *Radio Times*, a viewer from Romford in East London said:

> Like everyone else who witnessed the televised horror in the United States, I felt shattered and drained by the awful events unfolding before my disbelieving eyes. Brief respite came when my husband and I settled down to discover *The Blue Planet*. Surely there could be no more poignant contrast between the

evil that man is capable of perpetrating and the power, spectacle and sheer beauty of nature in all its spectacular glory on our precious Earth? Such compelling and beautiful photography was an absolute joy.

(Burgess, 2001)

Here then, the juxtaposition of images of real horror on television for much of the 30 hours of broadcasting that preceded it, and the spectacle and visual pleasure of *The Blue Planet* served, unintentionally of course, to emphasise the public service imperatives of the production of beautiful television. In the intervening decade, the BBC's blue-chip natural history programming has continued to be lauded as the 'best of television', specifically offering the viewer 'value for licence fee' in terms of both the extensiveness of its coverage of the natural world and in the visual and aural pleasure which it provides for its viewers. Producer Alastair Fothergill's fifth collaboration with David Attenborough, *Frozen Planet* (2011), broadcast ten years after *The Blue Planet*, echoes the superlatives applied in its reception. As with the latter series, letters to the *Radio Times* extolled the programme's dual delivery of education and entertainment: 'Thank you for *Frozen Planet*. This is what I pay my licence fee for and it's worth every penny. David Attenborough as always delivers with knowledge and flair, and the camerawork could not be excelled' (Brafield, 2011). The beauty of the camerawork and the particular suitability of this kind of programming to new high definition broadcast and reception were also commented on in the reviews for this series. As Robert Lloyd noted in the *LA Times*:

> The first thing to say about *Frozen Planet* [...] is that it is gorgeous to behold: lump-in-throat, tear-in-eye beautiful. It is the very point of such documentaries to be beautiful, of course, and not merely to honour, record and convey the awesome majesty of the natural world but also to look good on that big, expensive television set you bought yourself for Christmas.
>
> (2012)

However, other reviewers reported suffering from 'beauty fatigue' (Chivers, 2011), a feeling of being worn out by the spectacle of *Frozen Planet* and an overemphasis on visual pleasure:

> Lovely momentary images abound: a seal appearing in a wave like a ghost, seeming to float above the heads of the penguins watching it from the shore; the Antarctic sea ice, looking like a colourful abstract painting because of dead marine creatures pinned underneath it. These images will stick with you. But so will an overall sense that *Frozen Planet* is more – a lot more – of the same: an aestheticized, sentimentalized, anthropomorphic abstraction of the natural world in which gentle soundtrack music, winsome narration […] and the judicious use of slow motion combine to put us in a pleasant stupor on the couch.
> (Hale, 2012)

Sam Wollaston in the *Guardian* was more resigned to 'giving himself over' to this visual pleasure: '*Frozen Planet* is fabulous, beautiful, sumptuous […] So it's televisual wallpaper, so what? Wallpaper doesn't get much better. That's what this is about – a visual feast' (2011). These criticisms call to mind Shaftesbury's exploration of beauty in the eighteenth century and his contention that beauty was ultimately tied to a form of 'disinterested pleasure' (McMahon, 2005: 314), from which followed Kant's notion of the 'restful contemplation' of beauty ([1790] 1987: 101). Erica Carter, in her exploration of the sublime and the beautiful in Third Reich film, summarises that '[the Kantian conception of] beauty involves a refusal of the senses, and requires instead a disinterested mode of aesthetic contemplation' (2011: 142). In light of this, extreme close-ups of snow on fur emphasising texture and light, or the magnified formation of ice crystals, or long elegiac montages of melting ice accompanied by the ubiquitous soaring orchestral score[4] might be seen to provide a spectacle for the 'disinterested' or 'restful' viewer according to the above reviewers, a spectacle which is in keeping with this eighteenth-century conception of beauty. These theorists of beauty suggest that we languorously 'take in' the spectacle of the beauty of the natural world without becoming so overwhelmed or absorbed in it that we lose our senses (as we might in the face of the sublime). To scholars of television this description of the reception of beauty will, of course, recall broader and long-standing (though contested) critical depictions of the viewer of television who is interested, not absorbed, engaged, but in a distracted way.

This recourse to eighteenth-century philosophy also raises an interesting critical question: can we call this television programme 'beautiful' in itself, or is it simply showing us beautiful things?[5] Is the television programme a beautiful artefact or is it simply a conduit to something we might call 'the beauty in nature'? These questions recall Hall and Whannel's early discussion of television as both medium (i.e. a thing in itself) and a 'channel' (i.e. something that 'channels' other cultural forms) (1964); in this case, what it channels is nature, and specifically the beauty of nature. Kant considered nature 'on an equal footing with the fine arts' (Mothersill, 1992: 48) and his *Critique of Judgement* (1987), first published in 1790, in fact pays more attention to the beauty in nature than to other forms of the beautiful, looking back to earlier conceptions of beauty as encapsulating microcosmically the beauty of the universe in, for example, the work of Pythagoras.[6] In light of this, we wonder whether the reviewer above who finds *Frozen Planet* 'lump-in-throat, tear-in-eye beautiful' (Lloyd, 2012) speaks of the beauty of the *subject* of the series, the beauty of the natural world, or of the series itself, its camerawork, its combination of music and image, and so on. 'Beautiful' was used frequently as a nomination to describe the aesthetic of *The Blue Planet*. However, this concept was not fully explored, but rather its vernacular usage was borrowed, extended out from the analysis of the programme's reception in the article, rather than critically examined. The conclusion of this chapter will thus engage further with the extensive critical history of this term in order to consider what is at stake in calling television beautiful.

It is unsurprising that an exploration of beauty would turn back to the eighteenth century; in much of the twentieth and twenty-first centuries, the category of the beautiful became critically unfashionable, tied as it had been to assumptions about value which were dictated by class, race, gender and sexuality. Describing Bourdieu's approach to beauty, for example, Armstrong explains that he 'sees a concern with beauty, elegance and taste as performing a political function: establishing a set of values according to which the dominant class automatically comes out on top' (2005: 98). Thus Bourdieu (1984) associated notions of exclusion and elitism with the idea of beauty. Then there was the problem of pleasure. As Gaskell notes, 'Part of the heritage of modernism is its disavowal of pleasure. This led to a distrust of the notion of the beautiful for much of the twentieth century, a distrust ascribed upon the agenda of modernism' (2003: 272). The question of pleasure is explored at greater length below, but in relation to the natural history programme, it has been argued that beauty or spectacle might be distracting us from thinking about the harsher realities of the natural world (see Bousé, 2000, for example), and there is a critical tension implied in those arguments in which beautiful nature might simultaneously act as a distraction from pressing ecological issues just as these programmes show what is at stake in the loss of particular natural habitats (the loss of beautiful nature itself). Depicting the beauty of the natural world on television has, in the past, distracted from presenting the realities of the people who inhabit a beautiful landscape (in the case discussed in the previous chapter, the realities of the lives of indigenous Kenyans under colonial rule), although such beautiful scenes might also be the backdrop for more careful consideration of the social impact of the presentation or preservation of such a view. These 'beauty as distraction' arguments are summed up by Elaine Scarry in her proposal that 'beauty, by preoccupying our attention, distracts attention from wrong social arrangements. It makes us inattentive, and therefore eventually indifferent, to the project of bringing about arrangements that are just' (2009: 36). Extending this idea to the study of television more broadly, John Corner has also argued that in some areas of television scholarship, 'the pleasures of television may be seen as diversions from the serious and the source of various kinds of distorted or inadequate perceptions of the world' (1999: 94). Here Corner refers to texts such as Neil Postman's anti-television polemic, *Amusing Ourselves to Death* (1986), which positions the medium in general as a distraction from the political and social realities of 'real life', or those discourses that circulate in some areas of the press depicting television as 'mind-numbing' or 'hypnotic' in some way. The aim of this book is thus to closely examine such claims about television's *visual pleasures* and how they operate in relation to the close reading of particular texts.

The other key reason that the concept of beauty had been problematised in the twentieth century was in relation to its endlessly subjective definition. Mary Mothersill, one of the key contemporary theorists of beauty, issues this warning about the debates that surround this term:

> Few would deny [the] importance [of beauty as a topic of enquiry] and yet the mere suggestion that it be defined drives intelligent people to witless babble. They suppose that the first

and obvious requirement is to prove that beauty is 'objective' [...] The suggestion is that until what is assumed to be impossible has been achieved, there is no point in talking about beauty. But all this is pretentious nonsense.

(1992: 44)

Mothersill thus rather bluntly acknowledges the complexity and obfuscatory nature of the debate surrounding beauty, and particularly the critical obsession with the provision of objective criteria for its definition. Despite the work of artists and philosophers in the eighteenth century who sought to define beauty by a particular set of aesthetic criteria (e.g., Burke's *A Philosophical Enquiry into the Origin of our Ideas of the Sublime and the Beautiful* ([1757] 1990) and his delineation of the smallness, smoothness, fragility of beauty, or William Hogarth's *The Analysis of Beauty* ([1753] 2009) and its argument about the centrality of the serpentine line in the definition of the beautiful) no objective criteria for judging or defining the beautiful in any form or media has been (or can be) conclusively found. There are echoes here with debates that are central to the scholarship on television, and particularly the scholarship that seeks to define or analyse notions of 'quality television' or to make rigid evaluative judgements about the medium across the full range of programming that television has to offer. In Jennifer A. McMahon's rejection of the idea of the 'objective beauty checklist', for example, one hears echoes of Jason Jacobs' (2001) and Christine Geraghty's (2003) debate about quality television and the need for multiple sets of criteria by which to judge different types of television:

> One cannot predict that the presence of any combination of base properties (whether aesthetic or non-aesthetic) will give rise to the experience of beauty [...] There are no features that can be inducted over a number of cases of beauty to serve as sufficient conditions of beauty, nor can the features of one beautiful object be generalised to account for all cases of beauty.
> (McMahon, 2005: 309)

What this means for the study of television is that it would be as problematic to identify a singular, objective criteria for defining 'beautiful television' as it would to identify a singular, objective set of criteria by which to judge 'quality' on television. Whilst Brunsdon's work (1997) on quality in the early 1990s, discussed above, has often been misunderstood as providing such a singular evaluative criteria, what she in fact provides us with is a sketch of the criteria circulating around a particular set of texts at a particular period in time, and in doing so acknowledges the subjective (classed) position from which these criteria are developed.

Whilst the extracts of the article reprinted above show an engagement with the notion of beauty as it related to a particular programme, this offers only one, genre-specific version of what 'beautiful television' is or might be. In discussing the discourse of beauty and spectacle that surrounded *The Blue Planet* (in reviews, discussions by policy makers and producers and in the words of the programme's viewers) I implicitly argued that the definition of beauty was not solely the subjective judgement of the textual analyst, but rather a collective judgement which drew on a particular set of classed taste codes, taking my lead from Brunsdon's work on 'quality television'. Subsequently, in order to broaden and challenge our definition and understanding of beautiful television, and to contemporise it, I conducted a simple online survey that asked its respondents a single question: 'Are there any television programmes, past or present, that you would describe as beautiful? If so, please name them.'[7] By posing such an open question, respondents were invited to interpret the term themselves, and patterns and repetitions in these responses would subsequently become visible.

What was immediately striking in the responses to this question was that instead of featuring documentary, and particularly natural history, or more broadly those programmes which showed us 'beautiful things' from the natural world, it was drama programmes, and particularly 'high-end' crime dramas, that were the most oft-cited examples of beautiful television. The most frequently cited programmes were largely contemporary, reflecting the fact that those responding to the survey were easily able to recall beautiful television from their current viewing, though this was not true across the board; the most frequently cited example was broadcast over two decades previously. The ten programmes most frequently mentioned were:

1. *Twin Peaks* (ABC, 1990–1) 16.7 per cent of respondents mentioned this programme;
2. *Mad Men* (AMC, 2007–15) 13.6 per cent;

3. *Breaking Bad* (AMC, 2008–13) 8.6 per cent;
3. *Hannibal* (NBC, 2013–15) 8.6 per cent;
5. *Top of the Lake* (BBC, 2013–) 7.4 per cent;
6. *Utopia* (Channel 4, 2013–) 6.8 per cent;
6. *True Detective* (HBO, 2014–) 6.8 per cent;
8. *Les Revenants* (Canal+, 2012–) 6.2 per cent;
9. *Game of Thrones* (HBO, 2011–) 5.6 per cent;
10. *Planet Earth* (BBC, 2006–8) 4.9 per cent.

It might be surmised from this list then that in the contemporary moment, beautiful television is commonly understood to mean, primarily, big budget primetime drama, much of it imported from the US. We might also be able to build a tentative set of criteria for defining the beautiful on television from this list: for example, the dramas here are all distinguished by complex production design, all programmes are shot on location using HD or 35mm cinematography which emphasises the presentation of landscape as well as action on a more human (or animal) scale, and all feature extra-diegetic music which has an intense and meaningful relationship to the images on screen. This latter category seems to be particularly important, echoes my earlier analysis of *The Blue Planet* and its use of music, and can be related to British writer and philosopher John Armstrong's point about the beauty in 'composition'. Discussing Tony Palmers' 1983 film, *Wagner*, he says:

> Some of the most beautiful things we encounter involve the coming together of two apparently very different things. Words and music come together perfectly in some songs. Music and image are ideally suited to one another in the most attractive of films [...] But why should this be [...]? The music draws us into a particular emotional awareness, we are more alert, and responsive, to the appeal of the [images on screen] [...] In turn the image enhances our enjoyment of the music [...] visual features provide just the right emotional atmosphere within which to enjoy the charm of the music.
>
> (2005: 44–5)

The composition of beauty as outlined here by Armstrong is exemplified in this list by the combination of Angelo Badalamenti's mournful score and

Frank Byers' rich, mist-soaked landscape cinematography in *Twin Peaks*. Whilst Raymond Williams urged us to turn off the sound of our television sets to experience beauty (1974: 77), it is that combination of sound and image which is judged to have a 'beautiful effect' in the responses given to this questionnaire.

Whilst *Mad Men* offers visual pleasures which are tied to the careful recreation of a particularly 'stylish' period in US urban history (the pleasures of small details in heritage drama as defined by John Caughie in his discussion of this generic category of drama in the UK (2000)), the other drama series in this list all frequently feature characters exploring a landscape, searching for clues within that landscape, and often, simultaneously, some sense of existential meaning from it. This kind of narrative arc calls for careful composition and framing of (often beautiful) natural space; it also frequently features dialogue-less montages, accompanied by a non-diegetic score, of characters within this landscape.[8] This list also raises interesting questions about the ways that visual pleasure on television might either be tied to or separated from feelings of fear or unease that are inspired by the narratives of these fictional programmes. Eight out of ten of these programmes have a murder (or multiple murders) at the heart of their narratives; six out of ten feature sexual violence. Several of these series (*Hannibal*, *True Detective*, *Game of Thrones*, for example) feature extended shots or sequences of the tortured or mutilated body posed as if works of art (see Fig. 4.1). The beautiful as identified in this list can be related to that long history of works of art and culture which aestheticise the dead or decaying body as outlined by Elisabeth Bronfen (1992).

However, it would be problematic to either argue that this research has produced a definitive list of beautiful television, or an objective set of criteria by which we could define or judge beauty on this medium. Even within the above 'top ten' it is impossible to identify any meaningful features that are present across all of these examples of 'beautiful television', except perhaps that they were all (at least partly) filmed on location, but even this doesn't hold true if we look even slightly further down the list produced through this piece of viewer research. Whilst the most oft-cited categories of drama in the survey were the crime drama and the costume drama (which made up 28 per cent and 17 per cent of the examples given respectively), respondents also frequently mentioned, for example, other drama series, serials and

Visual pleasure, natural history television and televisual beauty

Fig. 4.1 The artistically mutilated body: *Hannibal* (NBC, 2013–15)

one-off teleplays, documentaries (particularly, though not exclusively, wildlife programmes), sport, reality television, children's television (e.g. *Bagpuss*, BBC1, 1974; *Abney and Teal*, Cbeebies, 2011–12), situation comedy (e.g. *'Allo 'Allo*, BBC1, 1982–92; *Only Fools and Horses*, BBC1, 1981–2003) and soap opera (e.g. *Coronation Street*, ITV, 1960–; *Eastenders*, BBC1, 1985–). Whilst the latter two categories might seem particularly surprising, we might relate their evocation to Roger Scruton's argument, via Rembrandt, that 'beauty is an ordinary, everyday kind of thing. It lies all around us. We need only the eyes to see it and the hearts to feel it'.[9]

The list of 'beautiful television' created by this research became so extensive and broad ranging that it served to confirm that (a) the categorisation of the beautiful is entirely subjective; (b) the beautiful is often understood to be synonymous with the pleasurable (see Mothersill, 1992: 46), and that it is sometimes hard to distinguish between the two; and (c) that beauty is found by viewers of television every day, across the broadcast spectrum and in almost every kind of programming and in every era of broadcasting.[10] This research also shows, as Gaskell points out, that over time and through its colloquial use, the term 'beautiful' has become 'no more than a general and vague term of approbation' (2003: 274), or, as Max J. Friedlander suggests, the beautiful 'suffers from an ominous generalness and painful vacuity' (1960: 87), but arguably it also tells us something beyond the fact that beautiful television is a vague and amorphous category in viewer discourse. What this research does is challenge the historically contingent work of those who rejected the concept of the beautiful on the grounds that it was a bourgeois or elitist category (see the discussion of Bourdieu (1984) above) by revealing the democratisation of the beautiful as a critical category of television for ordinary viewers. Even if in the equation of beauty and pleasure 'there is a special way in which we can become confused about beauty' (Armstrong, 2005: 67), the broader equation of beauty with *visual pleasure* in the above research touches on all of the ways in which we might find television visually pleasurable, from the 'lump-in-throat, tear-in-eye' (Lloyd, 2012) experience of the spectacular to the familiar and comforting sight of the places and spaces and people of television's more quotidian programming, or in our nostalgia for programmes past. Here beauty is understood as both a pleasurable treat provided by extraordinary television and as part of the ordinary, everyday pleasures of the medium, with visual pleasure repeatedly expressed as being part of the quotidian viewing experience of television.

Several respondents in this study also commented that whilst they struggled to assign the 'beautiful' label to particular programmes, they frequently encountered beautiful *moments* within television programming: one viewer commented that 'I think it is more common to find beautiful "moments" than it is to find beautiful shows' (29-year-old part-time teacher, London), whilst another eloquently argued:

> Many programmes – dramas, documentaries of all kinds – even news and current affairs, have moments or flashes of beauty – an image, a camera shot, a powerful use of audio or a moving dramatic moment or turn of speech; however, because any individual programme is composed of so many different elements I don't think I'd ever describe a whole programme as beautiful as I might a particular work of art. (57-year-old lecturer, Lincolnshire)[11]

The idea of beautiful moments or flashes of beauty makes sense in relation to a medium which is formally so hybrid and heterogeneous, for as many have argued (including Geraghty, 2003; Hilmes, 2008; and Jacobs, 2001) it is impossible to generalise about television aesthetics in a meaningful way across the entirety of the medium's output. Thus we might categorise television as a medium which is punctuated with beauty just as, in the final chapter of this book, the notion of erotic intermittence is employed to describe the presence of fleeting moments of desire on screen. This simple idea is, in fact, rather radical in relation to a medium which has often been uniformly associated with ugliness or at least a sense of relative aesthetic impoverishment in comparison to film and other visual media. Thinking back to the earlier analysis of *The Blue Planet*, and the suggestion that television's 'quality' needs to be assessed by looking at television 'flow', this idea can be extended to argue that moments of beauty routinely punctuate this flow, and that television must therefore be understood as frequently purveying the beautiful, or bringing the beautiful into the lives of its viewers. In light of this argument, the presentation of beauty might be seen as one of the many functions of a medium of visual and narrative pleasure, rather than as the antithesis of television's quintessential aesthetic.

So where does this leave us? In 2004 I argued that notions of visual pleasure were implicit in the discourse surrounding quality television and public service broadcasting in the UK, and that we should further examine the role of aesthetics in this debate. What a return to this question has shown is that the idea of beauty is still, and perhaps increasingly, important to viewers and the way that they evaluate programmes, and that the ways and places in which viewers experience beauty on television are diverse. Whilst the 'top ten' list reproduced above shows a perhaps unsurprising emphasis on (mainly recent) US long-form drama, if we look across the responses as a whole, we see that an overwhelming number of the programmes mentioned were produced by the BBC, the UK's publicly funded broadcaster with the most visible and significant remit for the provision of public service television. Even those responding to the survey from outside the UK most frequently cited BBC programmes in the category of the beautiful; there were over four times as many mentions of BBC programmes as those of ITV, and over six times as many as Channel 4 in the UK or HBO in the US. It is thus clear, even from this small-scale study, that the BBC is frequently considered to be providing beautiful programmes or moments of beauty within their programming; from this it might be deduced that there is something about the way that this broadcaster is funded and focused that produces beautiful television.

In light of this, the opening of *Frozen Planet* is particularly poignant. The series begins with a montage of sequences showing a variety of frozen landscapes, accompanied by David Attenborough's off-camera voiceover remarking on these images as both a 'magical' spectacle and a depiction of a wild environment at risk: 'These are places that feed our imaginations, places that seem to be borrowed from fairy tales [...] This is our planet's last true wilderness, and one that is changing, just as we're beginning to understand it'. Following this sequence, Attenborough appears on screen, standing in a frozen landscape where he continues his narration to camera:

> In this series we'll be travelling to all parts of these lonely lands, both north and south, to witness its wonders perhaps for the last time, and to discover some extraordinary examples of survival against the odds to be found anywhere on the planet.

The sequence then closes with an extremely long aerial shot of Attenborough standing alone on the icy crest of a mountain, which pulls up and away from him to emphasise his remoteness (he becomes nothing but a tiny red dot on the landscape) and to provide a space in the shot for the text 'to the ends of the earth', the episode title, to appear in the sky above him (see Fig. 4.2). Arguably, this sequence might be understood as triply elegiac. Firstly, Attenborough directly addresses the impact of global warming and the imminent loss of many of the planet's frozen environments in his on-camera narration when he speaks of witnessing its 'wonders perhaps for the last time'; this addresses the idea that the visual pleasure of this particular genre is also tied to showing us what is at stake in the damage which has been done to our environment. Secondly, as a solitary figure on the edge of an inhospitable environment, Attenborough's words recall the fact that at 85, this veteran broadcaster is nearing the end of his career, or almost certainly reaching an age where experiencing this spectacular view firsthand will become increasingly difficult; thus the sense of elegy is implicitly, if not explicitly, personal here. As the camera pulls up and away from him, the fragility of his position on the mountain is emphasised and invites us to contemplate Attenborough as an aging figure, even as he recedes out of view. Finally, the elegiac nature of this short sequence might also be tied to the position of the BBC as a publicly funded producer of television in a changing media landscape. As Fothergill

Visual pleasure, natural history television and televisual beauty

Fig. 4.2 An elegy for beautiful television? *Frozen Planet* (BBC1, 2011)

and Attenborough's programmes always emphasise, the BBC is figured as having gone 'to the ends of the earth' for the viewer; in this series and in others, the short 'making of' sequences at the end of each episode when screened in the UK stress the enormous temporal and financial undertaking involved in the production of the programme. In order to continue to make programming on this scale, the BBC relies on the maintenance of its funding structures and thus we stand to see this kind of programming 'for the last time' if this is not the case. Implicitly then, the spectacle of this sequence, and its image of beautiful nature, are what the viewer stands to lose if we don't value and work to protect the natural environment, or we don't value and work to protect the BBC as a national broadcaster. Here then 'beauty' on television is multifunctional, and the depiction of landscape is both visually pleasurable *and* meaningful in a number of ways. Whilst much of *Frozen Planet* can be considered beautiful, prompting the 'beauty fatigue' (Chivers, 2011) identified by its reviewers, when taken as a part of the larger broadcast 'supertext' (Browne, 1984), its episodes can be understood as providing significant moments of beauty within television's broader flow. Sometimes these moments are fleeting – a single shot, or sequence, a scene or setting – but they are inextricably woven into the fabric of a wide variety of television's genres.

5

Television's landscapes, (tele)visual pleasure and the imagined elsewhere

In the last ten years, there has been an explosion of rural imagery on British television, a veritable feast of rolling hills and dramatic coastlines, all accompanied by soaring orchestral scores and state-of-the-art HD aerial photography. Felix Thompson struggled with the categorisation of this programming as a 'geography genre', 'country themed series' and 'travelogue', ultimately proposing that 'it is more appropriate to think of a general geographical disposition adopted by British television towards the hinterland' (2010a: 65). However, one can better understand programmes such as *Bird's Eye View* (BBC, 1969–71), *A Picture of Britain* (BBC, 2005), *Coast* (The Open University/BBC, 2005–), *Britain's Favourite View* (ITV, 2007) and the *Wainwright Walks* series (Skyworks for BBC4, 2007–9) as forming a popular mode of landscape programming as well as a particularly spectacular mode of television documentary. Drawing on the literature on landscape in film and fine art, this chapter sketches out the identifying characteristics of landscape as it is presented on British television.

As a genre of painting, landscape art has been repeatedly understood (and often disparaged) as 'mass art' from within the academy; from the fifteenth century onwards, landscape painting was seen as less challenging than other forms of art, producing pictures that 'merely pleased the eye and provided little to occupy the mind', according to art historian

Norbert Wolf (2008: 9), though as Wolf also noted, 'the immense and growing public popularity of landscape pictures over the centuries gradually undermined [this] academic dogma' (ibid.: 11). It is this mass appeal in the image of the landscape that programme-makers across a number of televisual genres have attempted to harness, albeit to different effects. We frequently see extended montages of landscape on television in genres as diverse as the heritage drama,[1] natural history programming,[2] the cookery programme and children's television. Here then, presentation of televisual landscape in two kinds of programming will be explored: firstly, the landscape programmes described in the chapter's introduction, and, secondly, the holiday programme, in which landscape is a key signifier of an imagined elsewhere presented to the viewer-tourist. Across both case studies, the idea of visual pleasure on non-fictional television will be explored and related to the idea of a contemplative mode of viewing more traditionally associated with the spectacular in other media and at odds with both theories of the distracted viewer identified by early theorists of television (e.g. Ellis, 1982) and counter-theories of 'sit forward' viewer engagement or enthralment (developed particularly in relation to describing the viewing of 'quality drama' and in relation to Caldwell's notion of 'televisuality', 1995).

The landscape programme and the contemplative viewer

My initial interest in television's spectacular landscapes was prompted by three realisations (attached to three events in my personal and professional life) that provide an illuminating starting point for this analysis. The first was a visit to the BBC headquarters at White City at the end of October 2006. This date is significant because it comes four months after the beginning of the BBC's HDTV trial and coincided with the broadcast of the second series of *Coast* (the first season of *Coast* to use HD filming for its aerial photography, described on its DVD box set as 'stunning aerial images shot in High Definition – so you can see our beautiful coastline in unparalleled detail and clarity'). The lobby where I waited for my meeting to start was the first place I'd ever seen an HD television, here a large, expensive, flat screen set which faced the front doors of the reception at

White City, showing an HD showreel (which included images from *Coast* Series 2) accompanied by a voiceover and integrated text announcing HD television as 'the next big adventure' for the BBC, and pronouncing it 'the future of television'. I was dazzled, indeed, by the beauty and the clarity of the landscape photography, and it became clear to me that the programmes dwelling on landscape imagery on screen (*Coast, Planet Earth,* BBC/BR/WDR, 2006, and even the BBC's adaptation of *Bleak House,* Deep Indigo/Smallweed for BBC/WGBH, 2005) must have been made specifically with these new TV sets and their viewers in mind. This first realisation about television's spectacular landscapes related then to the complexity of the relationship between technological and generic change. All of the genres represented here pre-existed the advent of HD broadcasting in one form or another, and, in the case of the natural history programme and the literary adaptation, had also been associated with visual pleasure in earlier scholarly writing,[3] but it was also clear that certain spectacular forms and genres were being brought to the fore in a moment when an expensive new technology needed to be sold to the masses.

The second time these programmes came to my attention was during August 2007, when I was struck with the realisation that programmes featuring rural imagery had reached a point of critical mass, saturating the television schedule. For example, on Sunday 12 August, an impressive 33 per cent of the 8 pm to 1 am schedule on BBC1, BBC2 and BBC4 was taken up by what might loosely be described as programming which dwelled on rural landscape, alongside *Britain's Favourite View* on ITV1, scheduled against reruns of *Coast* on BBC2, and *Vicar's Wives* (KMB for ITV, 2007), set in the rural south coast of England and scheduled in a late night graveyard slot against reruns of *The Great British Village Show* (BBC, 2007) on BBC1. Clearly riding on the success of *Coast* and similar shows, the summer schedules featured a glut of programmes that dwelt on landscape imagery (specifically the landscapes of the British countryside at the time), and in which the art of walking (or rambling, sauntering or strolling) through these rural spaces was often matched with the search for a 'perfect view'. For at least part of each of these programmes (and others beyond) we saw what Martin Lefebvre, in his analysis of landscape in cinema, refers to as 'space freed from eventhood' (2006a: 22) (as opposed to land as setting). This wasn't just Sunday night programming, however; it

Television's landscapes, (tele)visual pleasure and the imagined elsewhere

Fig. 5.1 Television reconfigured as wall-hanging art

was everywhere, throughout the week, creating a nostalgic, 'heritage' image of 'unspoilt Britain' entirely devoid of urban space or industrialisation. As I looked forwards and backwards across the TV listings guide, it became clear that a cross-generic mode of programming had developed – the landscape mode – which had an aesthetic history within and beyond television, a mode that challenged many of the critical assumptions about television's lack of interest in, and inability to produce, the spectacular.

The third realisation relates to the moment in 2009 when the first member of my family bought an HD ready, wall hanging, flat screen TV. This set (see Fig. 5.1), hanging on the wall in my brother-in-law's house, clearly had the spatial characteristics of a painting, and was positioned in the room where a large print had once been. Debbie Rodan's analysis of the marketing of these new sets stresses, in particular, the ways in which they have been positioned by those selling them as objects of 'aesthetic beauty' whereby the 'screen is rendered spectacular, visually pleasing and placed on display' (2009: 371). She goes on to note that 'the architecture of the room [in the advertising of these sets] is signified as the interior of a public gallery space' (ibid.). Considering Rodan's analysis in relation to my brother-in-law's television set, I began to look for other images of the positioning of these new sets and came across, firstly, the 'TV frame',[4] where an add-on frame could be purchased to surround a flat screen set, giving it the physical appearance of a piece of art rather than a piece of domestic technology within the living room, and, secondly, the phenomenon of the 'art screen TV', in which the new flat screen HD TV could be framed as a piece of fine art might be *and* hidden out of sight when not in use by a reproduction painting (often a landscape painting) which could be drawn up and down.[5] Here, television as an object was being given the visual status of fine art within the living room, and the predominance of landscape imagery on TV (and sometimes over it) began to make more sense in relation to this; television's current wealth of landscape images might in fact be partly understood as television being occasionally viewed and appreciated as popular, 'ambient' art. The sleek, minimalist, galleryesque rooms, identified by Rodan and also seen in the promotion of the products discussed above, are, therefore, the (real and imagined) domestic spaces of HD television.

This contemplative mode of viewing is clearly also encouraged by the repositioning of the HD, flat screen television set discussed above; the shift in spatial relations in the contemporary living room may indeed partly invite the recent flurry of landscape programming. Whilst rejecting a technologically determinist argument about the rise of HD shooting and viewing technologies and the advent of this mode of programming (earlier examples of such programming suggest the mode has a much longer history, as discussed below), it can also be argued that the current cycle of programmes, proliferating since 2006 as at no other time in the genre's history, must be understood as post-digital revolution television. This is simultaneously 'slow television',[6] which allows for a contemplative gaze on spectacular 'natural' landscapes, and also a heavily CGI'd cycle of programming which draws on a 'Google Earth' aesthetic through the pronounced use of satellite imagery to produce a frenzy of dazzling topography, showcasing the spectacle of such 'new' technologies.

Making this argument is not to propose that the presentation of landscape has never been offered on television before: we could look at much earlier examples of landscape programming to locate an emphasis on spectacle and visual pleasure prior to the advent of HD broadcasting. For

Television's landscapes, (tele)visual pleasure and the imagined elsewhere

example *Bird's Eye View*, described by the BBC in a web-based retrospective of the 'aerial journey' programme as a 'series of helicopter travelogues providing breathtaking views of the British Isles',[7] was produced between 1969 and 1971 by Paul Bonner and series-edited by Edward Mirzoeff, with several episodes, including its opening one, scripted and narrated by the poet John Betjeman. The programme was shot entirely from an Alouette II helicopter by Geoff Mulligan, and although this earlier series lacks the image quality of later examples of the landscape mode on television, its emphasis on the spectacle of aerial photography, and on spectacular montage (a term defined at greater length below), very much aligns it with the visual pleasure of later series such as *Coast* and *A Picture of Britain*. Its title sequence, a jazz-scored montage of aerial filming in action which moves from close-ups of parts of the helicopter from which the series is filmed to much longer shots of the aircraft hovering above unidentified green fields, spectacularises and glamorises the means of production: the final shot of this sequence before the title comes up sees the helicopter fly into the direct line of the sun, bathing the machine in a kind of 'heavenly glow' and suggesting an omniscient viewpoint for the series. In contrast, in the first episode of the series, 'An Englishman's Home' (tx. 5/4/69), a slower montage of aerial footage of that great British signifier, the stately home, follows this title sequence. This montage, accompanied by the first line of John Betjeman's poetic voiceover ('There's a saying, you've heard it before, the Englishman's home is his castle. Well I suppose, in a way, it is') and Dame Nellie Melba singing Sir Henry Bishop and John Howard Payne's 1923 classic, 'Home, Sweet Home!', further emphasises an aestheticisation of the English landscape on television, through lingering aerial photography that moves slowly around a series of large houses set in rural spaces.[8] This series, whilst offering commentary which was sometimes highly critical of the human impact on the British landscape (as in the episode 'Green and Pleasant Land', tx. 22/6/69), can be seen as a precursor of the more recent series considered in this chapter, given that it similarly capitalises on the rural and coastal views of Britain in a way which emphasises a contemplative appreciation of this space as a 'view'.

The use of archival material *within* the contemporary landscape programme also reminds us that the spectacle of the national tour, the desire to access remote and beautiful places *for* the viewer, has been a constant on British television. For example, towards the beginning of the first episode of *A Picture of Britain* ('The Romantic North', tx. 5/6/05), highlights from presenter David Dimbleby's appearance (with his brother, Jonathan) on an episode of the UK travel series *No Passport* entitled 'The English Lakes' (tx. 24/2/60) are edited into his return to the same location. The sequence begins with a cut from Dimbleby in his car (in 2005), talking to camera about his first 'visit' to the area, to a shot of him driving a car past the camera on a road signed to Windermere in 1960. What follows is a montage of images from their programme on the Lakes, culminating in a picture-postcard shot of Ashness Bridge with Derwent Water and Skiddaw mountain in the rear of the shot, slowly dissolving from black and white 1960s footage to sumptuous colour footage taken from exactly the same camera position in 2005. The slow dissolve here means that the continuity of television's landscape imagery is emphasised through a 'ghostly' overlaying of footage from two eras, reinforced by Dimbleby's observation that 'the extraordinary thing is it really hasn't changed much'. Here, then, a piece of television history is inserted into the episode to assert three things: the unchanging nature of British television (albeit with Dimbleby later commenting on the shift in standardised accents for television), alongside the unchanging nature of the British landscape and, by association, British national identity. This sequence thus warns us of the problem of viewing television landscapes as a recent phenomenon, tied solely to the advent of HD broadcasting, whilst simultaneously alerting us to the rather jingoistic ways in which the representation of landscape has been, and continues to be, tied to an image of national identity as simple and unchanging.[9]

It perhaps doesn't need to be stated that one of the key strands to have emerged from theorisations of landscape, particularly in the field of cultural geography, has been its distinction from 'land' – as W. J. T. Mitchell puts it, 'Landscape is a natural scene mediated by culture' (1994: 5); Simon Schama has similarly argued that 'landscape is the work of the mind. Its scenery is built up as much from strata of memory as from layers of rock' (1995: 6). Cultural geographers have sought to understand the ways in which human interaction with nature and environment have produced landscape, and arguably we must see the programmes at the heart of this study as part of this act of cultural production; their handling of space and place *affirms* landscape's position within the national imaginary, even though these programmes often place narrative emphasis on 'discovering' rather than 'constructing' the natural scene.

In attempting to understand the popularity and pervasiveness of landscape subjects in fine art, it has often been argued, particularly from a Marxist position, that landscape art acts as a response to an increasingly urbanised or technologically mediated world. This is what Malcolm Andrews calls the 'compensation thesis' (1999: 18). As Norbert Wolf suggests, 'People to whom nature appears in the form of landscape no longer live unthinkingly in nature. They are alienated from it, and can feel one with nature only through the mediation of aesthetics' (2008: 8); at this point the aesthetic value of land replaces its 'use and dependency value' (Andrews, 1999: 21). Wolf points out that this is not a recently formulated position: 'In 1767, the French philosopher Denis Diderot said that landscape paintings were hung on the walls of salons by city dwellers in order to compensate for their loss of contact with nature' (2008: 8). From a television studies perspective, this is very similar to those 'television as window on the world' arguments – at their most developed and sophisticated in the work of William Boddy (2004) and Lynn Spigel (2001) – where television promises a fantasy of 'armchair travel', by bringing 'the outside world into the private home'. If this is so, we can understand the television set showing extended montages of landscape imagery as a form of *trompe l'oeil*, a faux-window from which the urban viewer may view/consume the rural. As Raymond Williams has argued, 'the very idea of landscape implies separation and observation' (1985: 126); the positioning of the HDTV viewer in the domestic space cum gallery emphasises this tension between proximity and distance in relation to landscape spectacle. Indeed, the use of intra-diegetic frames in ITV's 'vote for your favourite landscape' show, *Britain's Favourite View*, supports this reading of television's landscape images as *trompe l'oeil* (where we see the intersection of artwork and window). At the end of each segment of the programme, in which a celebrity talks about and 'takes in' a particular view, the view is framed by a semi-transparent, bevelled CGI frame as the culmination of the presenter's endeavours to persuade us of the image's status as Britain's favourite view, as the title of the view is also displayed diegetically. This is seen, for example, in the final shot of Des Lynam's championing of the Seven Sisters in Kent, led up to by a montage of landscape shots accompanied by a Elgar's 'Land of Hope and Glory' and Lynam reading Kipling's poem 'Sussex' (see Fig. 5.2).

Fig. 5.2 The framed view of intentional landscape: *Britain's Favourite View* (ITV, 2007)

This 'framing moment' thus leads to the centre of this analysis of landscape television; in this mode of programme-making, a variety of audio-visual strategies seek, at various points in the episode, to halt the narrative progression of programmes such as *Coast*, *Wainwright's Walks*, *A Picture of Britain* and *Britain's Favourite View* in order to appeal to a *contemplative viewer* who appreciates a televisual form of landscape spectacle on an aesthetic level. In the context of the advent of HDTV, these moments have been particularly emphasised and extended. Martin Lefebvre's work on landscape and film is extremely useful here, in his discussion of two possible modes of spectatorial activity in relation to the viewing of a single film: 'a *narrative* mode and a *spectacular* mode'. Lefebvre states,

> The contemplation of the setting [in the spectacular mode] frees it briefly from its narrative function (but perhaps, in some cases, only for the length of a thought); for one instant, the natural, outdoor setting [...] is considered in its own right, as landscape.
>
> (2006a: 29)

In the contemporary landscape factual entertainment programme then, narrative progression is frequently slowed or halted to enable contemplative viewing.

An example of this spectacular mode of contemplation can be seen in the first episode of *Wainwright Walks* produced by the aerial filmmaking company Skyworks for BBC4. This is an interesting programme for this study. The 30-minute episodes are each based around the presenter, Julia Bradbury, completing one of the walks from Alfred Wainwright's famous guidebooks to the Lake District; as such, it is a rather sparse, empty programme which punctuates sequences of Bradbury's walking (and talking to camera) with shots of the surrounding scenery. In episode one, on leaving Low Wax Knot on her ascent of the Haystacks, Bradbury exclaims, 'You wouldn't think the view could get any better [...] but it does'. The following sequence is a slow track down the hills, eventually showing an extreme wide shot of Bradbury sitting 'alone' within the landscape, accompanied by her voiceover describing Wainwright's passion for the landscape, and the voice of actor Nik Wood-Jones reading Wainwright's words in a rumbling Lancastrian burr ('the magical atmosphere of the lakes, the silence of lonely hills, the dawn chorus of birdsong, silver cascades, dancing and leaping down bracken steeps, and the symphonies of murmuring streams'). This camera movement finishes on a close-up of Bradbury gazing into the distance (held for a quiet five seconds), after which the shot moves out to a slightly wider profile shot, during which Bradbury extols the virtues of walking alone and the serenity and peacefulness of 'being alone' and having the luxury of 'all this space' (see Fig. 5.3). Whilst Bradbury's solitude is, of course, an illusion (the presence of the crew denied by her dialogue), the following repeated sequence of her sitting as a solitary figure in the landscape tracks off into another 'appreciative view' of the distant hills and is accompanied by a quiet crescendo of strings and piano. This combination of enraptured presenter, slowly moving tracking shot around the landscape and orchestral accompaniment all denote that this is, to borrow a term from Charlotte Brunsdon's work on cinematic 'empty spaces', a 'hesitation' (2007: 219) in the televisual image where the emptiness of landscape, or what Lefebvre calls 'space freed from eventhood' (2006a: 22), allows for moments of contemplation, or breaks from narrative progression, both when Bradbury herself is resting (as in this moment), and beyond. The

Fig. 5.3 A hesitation on intentional landscape: *Wainwright Walks* (Skyworks for BBC4, 2007)

on-screen presenter is of course critically important in the landscape programme, pointing the viewer towards what Lefebvre calls 'intentional landscape' (ibid.: 30), that is, spectacular imagery marked out as landscape by the programme-makers (rather than that which is discovered or understood by the viewer as landscape without any obvious framing of it as such). Julia Bradbury in *Wainwright Walks*, Nicholas Crane et al. in *Coast*, David Dimbleby in *A Picture of Britain*, Andrew Marr in *Britain from Above* (BBC1/BBC2/BBC4/BBCHD, 2008) and a plethora of celebrity presenters in *Britain's Favourite View* figure as what Lefebvre has seen in the fiction film as a 'character enraptured by the natural space offered to their gaze [that] can lead the spectator to contemplate the same space as an autonomous landscape' (ibid.: 33).

Given that many of these programmes feature presenters walking through or physically exploring land (and landscape), it is unsurprising that the adjectives one wishes to use to describe the camerawork here are associated with that activity: the camera lingers, it meanders and rambles over this space, inviting a contemplative gaze. This is 'slow television' for the contemplative viewer, to borrow Steven Poliakoff's phrase; indeed, executive producer of *Wainwright Walks*, Richard Mervyn, has been keen to stress that this slowness was part of the programme's appeal to an older audience, equating a faster pace with the youth programming of BBC3 as opposed to BBC4, the original home of his programme;[10] indeed, it is unsurprising that when the genre of slow TV came to the UK in 2015, it was located precisely on this channel. (See the Introduction of this book for a discussion of slow TV.)[11] If this interpretation is extended, this is a kind of 'screensaver TV' that privileges the prolonged montage of moving images of landscape set to orchestral music to a great degree.[12] In fact, across nearly all genres of programming, *montage* is to television spectacle what the long take and wide angle shot are to filmic spectacle; the extended montage without dialogue asks viewers to *look* rather than to follow narration. Sometimes this moment is also an invitation to the viewers to creatively make connections between scenes and images in order to tell a story, frequently at the end of an episode (for the sake of clarity, 'narrative montage'); however, at other times we are invited to simply look and enjoy, to sit back and relax, to contemplate a series of beautiful or striking images on screen (the 'spectacular montage').

In these landscape programmes then, spectacular montages punctuate every episode: they are montages of movement, travelling shots, which dwell on what Kant describes as the 'noble sublime', that version of the sublime which is accompanied by a sense of 'quiet wonder' rather than the 'certain dread or melancholy' of the terrifying sublime (1960: 47). This quiet wonder is perhaps rather more suited to cosy, *televisual* images of landscape. There is a sumptuousness and a rather sensual quality to this imagery, referred to by series producer Steve Evanson as 'landscape porn';[13] in an interview, Evanson acknowledged that the series oscillates between presenting the coastal landscape as setting (for the variety of social, natural and geographical histories told) and presenting landscape for purely visual viewing pleasure:

> Aerial photography is not only used for visual impact, stunning shots of Lulworth Cove or some amazing Beachy Head, you know, or somewhere where you just want to […] basically it's almost like kind of landscape porn, and you're just kind of looking at it going kind of 'Wow, look at that', but we also try and use the aerials also as embedded in the stories.

You can see this pull between narrative montage and spectacular montage throughout *Coast* where the aerial photography shifts (sometimes imperceptibly) between 'landscape porn' and landscape embedded in storytelling (or spectacular montage and narrative montage). Porn is a good point of comparison here, given the fact that the spectacular landscape imagery of *Coast* and other programmes in this cycle appears to teeter on the brink of the sensational and gratuitous.

The production economy of these programmes, in relation to the expensive aerial photography they rely so heavily on, is interesting for a number of reasons. Programmes such as *Coast* and *Britain's Favourite View* 'buy in' their aerials from companies such as Castle Air, Flying Pictures and Skyworks, and the production companies negotiate the rights to reuse the aerials beyond their original use. (The BBC owns the rights to *Coast's* aerials, produced by Castle Air, whilst Skyworks nearly always retains the rights on their footage, and make a lot of their revenue through archive sales, rather than the production of new programmes and footage.) This means that across television production, in as many genres as use aerial footage of landscape, an image repertoire of stock shots and stand-alone sequences has begun to develop, which means that the same spectacular montage may be seen time and again outside of its original broadcast context (footage from *Coast* has been used in programmes as diverse as the Christian hymn-focused programme *Songs of Praise*, BBC1, 1961– and daytime, wheeler-dealer bargain show *Flog It*, BBC2, 2002–, for example). Extracted from its narrative framework, the use of the aerial landscape montage sequence becomes more obviously intended for the production of visual pleasure, beyond its use as a shorthand device for establishing place. Even a programme like *Wainwright's Walks* that was commissioned by BBC4 from the aerial specialist production company Skyworks relies heavily on what we might term the recycling of spectacle (Richard Mervyn, the company's founding director, reported that Series One was made up of about 40 per cent of their stock archived footage, rather than aerials that had been shot specially for the programme). Both of the producer interviews conducted in relation to this research also emphasised the skill and creativity required of the small number of film pilots and specialist aerial camera operators working in this area: Richard Mervyn, for example, argued that 'in a sense the pilot is not just a pilot, he's the grip, the helicopter is the dolly, a crane and a tripod, so you must consider that the pilot is part of the creative team'. Given that a relatively small number of flying and shooting experts are working across a range programmes to produce a vast amount of footage, it is perhaps unsurprising that television's aerial landscapes, seen in a range of genres from cookery programmes to reality shows like *The Apprentice* (Mark Burnett Productions/Talkback for BBC, 2005–), are also marked by a certain visual similarity.

Aside from the spectacle of aerial photography that dominates this cycle of programmes, we might also locate the production of visual pleasure within HD television's spectacular landscape documentaries in the use of satellite and computer generated imagery. If, as a kind of middle-brow landscape art of constantly moving pictures, these programmes seek to represent the 'natural beauty' of land rendered as landscape, then they also feature a dazzling array of what we might call special effects or trick shots. Firstly, CGI is frequently used to draw further attention to the pictorial qualities of this branch of factual entertainment, as seen in the CGI 'frame' example from *Britain's Favourite View* discussed above or in the opening title sequence of *A Picture of Britain*, where a watercolour painting is 'magically' completed in order to draw our attention to the process by which the British view becomes landscape: as David Dimbleby looks on in the left-hand foreground and subsequently moves out of the shot as it re-frames, his view of some 'typically British' rolling hills is replaced by a painted image of the same scene which slowly 'appears' over the original image. Computer generated footage is also used to draw on the spectacular imagery of other media forms, specifically Hollywood cinema, to 'sex up' a potentially visually uninteresting story in the specific case of *Coast* (as seen in the huge scale of the CGI recreation of the Titanic in a story about Belfast shipbuilding in the first season episode on Northern Ireland, or the CGI tsunami created for the opening episode on Shetland for Season Three).

Thirdly, and more prominently, however, the programmes employ a 'Google Earth' aesthetic through the insertion of computer generated and satellite imagery, and in doing so make a visual point about the relationship between topography, satellite photography and landscape art. The handling of the map as an aesthetic object was explored most thoroughly in the BBC4 series *The Beauty of Maps* (2010); arguably this series was a direct descendent of the landscape programmes at the centre of this analysis, which frequently featured digitally created maps at the centre of their narratives, or at punctuating points throughout them, and which often handled the representation of cartography in explicitly spectacular ways. The most obvious examples of this are the vertiginous zooms into a locating point of the map of the British Isles that punctuate each season of *Coast*, and which also dominated *Britain from Above* (the opening episode of this series drew particularly heavily on satellite and CGI images of the landscape as map in tracking the movement of various forms of transport around the UK). In the example of the characteristic *Coast* 'zoom', a dazzling topographical image maps the British landscape into a visual system which proposes that new technologies of seeing and surveillance have land, rather than landscape, absolutely covered, and that 'land' (and also 'terrain', 'territory') has entered into and been processed by technology, a technology that 'stands for' or represents systems of knowledge and power in a rather Orwellian sense. Simultaneously though, this is a beautiful image which offers a moment of heightened and exhilarating visual pleasure, showcasing the possibilities of digital satellite photography, and drawing on the popularity of the 'Google Earth' image to locate one's place in the world (Google Earth, I would argue, has become an internet leisure/'play' application for the majority of its users). As geographers Martin Dodge and Chris Perkins argue:

> In the past, satellite imagery was mainly deployed as part of rational scientific discourse, to target enemy facilities, manage environments, monitor land-use change, or as evidential support for planning enforcement. Increasingly, however, different social groups are deploying high-resolution satellite imagery in new ways. Actors in the process now range from mass-media, to artistic practitioners [...] to everyday leisure users [...] The avowedly naturalistic look of the virtual globe shrouded in satellite imagery is beginning to replace the world map of nation-states as the default meta-geography of the media.
> (2009: 497)

Paul Kingsbury and John Paul Jones III go further in describing Google Earth specifically as 'a Dionysian entity, that is, the projection of an uncertain orb spangled with vertiginous paranoia, frenzied navigation, jubilatory dissolution, and intoxicating giddiness' (2009: 503). In *Coast*, the zoom into the map of Britain provides the viewer with an exhilarating rush, Kingsbury and Jones' 'intoxicating giddiness', as if the journey down from space forms the first step of the journey around Britain that is about to begin; it therefore has both narrative and spectacular function, and figures topography as a sensuous and aesthetic and *spectacular experience*. Here we see television's landscape art producing the same kind of 'topographical sensibility' which dominated Flemish landscape painting in the sixteenth century; the map is intertwined with a vision of 'landscape'.

It is obvious that there is a great deal more to say about these landscape programmes, not least about the ways in which they engage in the creation and romanticisation of a 'timeless', unchanging Britain and, by association, British national identity. Furthermore, the environmental conscience of some of these programmes (particularly *Coast* and *Britain from Above*), or the rather condescending way in which programmes like *A Picture of Britain* deal with the question of rural tourism and its environmental impact, seems at odds, to me, with their visualisation of what John Urry has called 'the tourist gaze' (1990);[14] they often look and sound like holiday programmes, and the economy of the 'view' is always an underlying issue in this programming which both expresses discomfort at the commodification of the British landscape whilst at the same time cashing in on this very thing. In fact, the rise of these programmes on British television coincides with the disappearance of the holiday programme from the prime evening slot in the British television schedules,[15] a fact which perhaps reflects the current economic climate of the UK and its broadcasters, as well as the middle classes' increased consciousness of the questionable ethics of international travel in relation to the issue of climate change. It is to this holiday programme that we now turn.

Holidays, the tourist gaze and the imagined elsewhere

Joshua Meyrowitz has argued that 'although we always sense the world in a local place, the people and things that we sense are not always local. Further although all experience is local, we do not always make sense of the world from a purely local perspective' (1989: 327), meaning that our horizons have been expanded by the presentation in the media of places beyond our immediate locale. Whilst the television landscapes discussed thus far in this chapter have, to a greater or lesser extent, been local, or at least 'national', television presents distinctly 'unlocal' landscapes via the holiday or travel programme: in short, the far away. Meyrowitz writes of a 'generalised elsewhere', a mediated place and space beyond our local existence which is juxtaposed with our specific viewing contexts via television; here we instead think about an 'imagined elsewhere', an alternative landscape which is as specific as the places from which we view, but which is imagined, brought into being by the programme-makers and traversed and viewed *on our behalf* by a series of different diegetic figures. Whilst a wide variety of programming genres, from documentary to the cookery programme, extend our knowledge and understanding of this 'imagined elsewhere', the remainder of this chapter is particularly interested in how the ways in which holidaymakers, tourists and travellers engage with landscape are visualised on screen.

A growing body of literature considers the relationship between tourism and the media; beyond John Urry's account of the 'tourist gaze' (1990), David Crouch, Rhona Jackson and Felix Thompson's wide-ranging collection on this subject proposes that 'there is an overarching and necessary interdependence between tourism and the media [...] The media are heavily involved in promoting an emotional disposition, coupled with imaginative and cognitive activity, which has the potential to be converted into tourist activity' (2005b: 1). For these authors, television is one aspect of the media which both creates expectation and desire around a particular site of tourism, and also teaches us what to do, and how to view, when we are in that place. Fred Inglis also proposes that 'television is the source of the imagery with which we do our imagining of the future, and the holiday imagery now so omnipresent on the screen [...] is one of the best places to find our fantasies of the free and fulfilled life' (2000: 5). Whilst Inglis proposes that the holiday on television offers an imagined future, this chapter proposes that it also offers us an imagined elsewhere and, more specifically, an imagined landscape.

The holiday programme has a long history on British television which extends back well before the advent of the BBC's long-running holiday programme, *Holiday*, in 1969[16] and its ITV equivalent, *Wish You Were Here?*, originally broadcast in 1974.[17] On BBC television the holiday programme initially developed out of afternoon programming made by the Women's Television Department where much of the consumer-focused programming had found its home. From the beginning of 1952, for example, the long-running magazine programme *Leisure and Pleasure* (BBC, 1951–5) frequently featured items on planning a holiday in the UK and abroad, as did the magazine programmes *Your Own Time* (BBC, 1955–8) with its recurring feature on 'Dream Holidays' and the magazine show for older women, *Twice Twenty* (BBC, 1955–8). Out of these features grew separate holiday programming, still presented in the 'Mainly for Women' afternoon slot, later in the 1950s: these included *Happy Holidays* (BBC, 1956–7) and a series of programmes produced by Monica Sims – *Holidays Ahead* (BBC, 1957–8), *Holidays* (BBC, 1959) and *Remembering Summer* (BBC, 1959–60). These programmes were mainly instructional, offering a series of 'tips' for travelling and booking holidays by experts from the field of travel journalism as well as the travel industry (alongside advice on holiday fashions and health care abroad), and spoke to their viewers as wary consumers. For the BBC, then, the consumer address of the holiday programme grew out of their public service remit to keep their viewers informed about the issues of the day, and this was extended into programming with a more masculine address when BBC2 began broadcasting: for example *Time Out* (BBC2, 1964–5), a programme about leisure pursuits presented by Ludovic Kennedy, or the long-running motoring programme, *Wheelbase* (BBC2, 1964–74), which frequently featured items on holiday motoring and motoring holidays. This programming particularly instructed viewers on the perils and possibilities of international travel at the point at which it began to become an affordable reality for much greater numbers of the British public. On ITV the holiday programme also developed out of consumer-focused television, though under a different, rather more commercial guise. It was the commercial 'Admag', described by Janet Thumim as advertising magazines in which 'several separate but related items [about consumer goods/choices] were linked by a presenter – or "host" – and/or through a consistent mise-en-scène' (2004: 39), that initially addressed the viewer-consumer shopping for holidays on ITV. Holidays featured in special holiday Admags, including *Where Shall We Go?* (ABC, 1957–61), *Over the Hills* (AR, 1957–63) and *Motoring Holiday* (AR, 1957–60), as well as recurring features in long-running Admags like *Jim's Inn* (AR, 1957–63), *Get This!* (ATV, 1959–62) and *Showcase* (ITV, 1960–2). In a *Jim's Inn* episode broadcast on 15 August 1962, for example, 'customers' Fred and Roma visited Jimmy and Maggie Hanley's pub (the eponymous inn) to discuss their recent holiday, thus weaving consumer advice about holidays abroad into the ongoing narrative of life in the fictional suburb of Wembleham. On the other hand, *Get This!* addressed the viewer-consumer more directly on 17 January 1960, in presenting 'exciting and romantic holiday resorts all over Europe', according to the *TV Times*; viewers could then write in to the production team for details of any of the holidays featured. When Admags were abolished in 1963 as a result of recommendations made in the report of the Pilkington Committee,[18] the holiday programme on commercial television receded from view for the next decade, and was only revived (as *Wish You Were Here?* in 1974) as a response to the popularity of the BBC's *Holiday* programme.

This early history is interesting because it tells a story of the thin line that the holiday programme had to continue to tread between information and entertainment, commercialism and the production of the informed consumer. It is interesting that this programming was left out of the BBC's documentary on the holiday programme, *The Way We Travelled* (BBC4, 2007), which chose instead to identify the starting point of this genre as the short-lived Richard Dimbleby series *Passport/No Passport* (BBC, 1960) and the career of veteran traveller-broadcaster Alan Whicker. Aside from the fact that the extant archive of episodes of the magazine programmes and Admags discussed above would be patchy, if not non-existent, the BBC may also have been further seeking to distance this genre of programming from its commercialist roots. When the corporation first established its long-running holiday programme *Holiday*, broadcast in 1969, it was keen that this series would be based on travel journalism and detailed research into the activities of the British holidaymaker. It had a strict policy disallowing its production personnel from receiving any favours from the travel industry (though files at the BBC's Written Archives Centre suggest these were offered), and were keen to provide balance and objective reporting in its presentation of the holiday destinations at hand. Extensive research was carried out by the BBC in anticipation of the programme, written up by producer Peter Bales, whose report reflected on the increase in package holidays in the late 1960s and also on the impact of the imposition of £50 maximum foreign travel allowance (limiting the amount of currency British citizens were allowed to take out of the country from November 1966): 'Forty per cent of continental holiday makers (just over two million) bought a package deal in 1966. The indication is that this figure will continue to rise. They are undoubtedly the "trendsetters" and should feature predominantly in the programmes' (Bales, 1967). The BBC was thus keen to produce a programme that would reflect the kinds of holidays their viewers were taking, and to examine, rather than influence, holiday trends. John Carter, then a travel journalist at *The Times*, was partly responsible for writing early scripts, and later for presenting on both *Holiday* and *Wish You Were Here*, as were other 'serious' travel writers, such as Peter Whelpton of the *Daily Mail* and David D. Tennant of Thompson Newspapers Ltd. These men thus brought to the programme the values of investigative journalism as well as extensive knowledge of the travel industry, and were presumably employed partly as a way to counter the threat of commercialism identified by BBC producer John Lloyd in his treatment for the series:

> [Any] information must be conveyed in a way which avoids any impression that the programme is merely a television travel brochure. This can be done by incorporating within the framework of the programme a strong element of challenge. In each programme, the country or region involved will be told that the onus is upon it, collectively, to persuade the audience that it is worthy of their consideration. The programme itself will act as a combination of prosecuting counsel and examining magistrate. The audience at home is the judge.
>
> (Lloyd, 1967)

Lloyd's reference here to putting the onus on the country or place under examination to persuade the viewer of its own merits may refer to the fact that the production team had initially thought that they would use stock footage supplied by regional or national tourist boards or agencies to present place/landscape on the programme. However, ultimately the crew ended up shooting the majority of the footage themselves, guided by itineraries set out by the above travel journalists. These production decisions were thus partly made in an attempt to balance the presentation of spectacular landscapes and views seen through the optic of the tourist gaze with an examination of the realities of each place as a holiday destination. Tom Savage, the producer of *Holiday*, wrote in his copy for a preview of the series in *The Listener* that presenting these images of holiday destinations on film, 'supported in the studio by facts and figures, and a resident team of experts, is an attempt to counter-balance the commercial claims of the glossy brochures and the host of dream-like advertisements that bombard the public at a time when the hard selling begins' (Savage, 1968). There was thus an anxiety that a beautiful view needed to be 'counter-balanced', not allowed to exceed its narrative value, on the holiday programme. Whether this was achieved, ultimately, is examined below. It is certain, however, that the holiday programme was popular and successful enough to survive over 30 years of broadcasting.

The relative decline of the holiday programme in the mid-2000s reflects a number of things including the slashing of programme budgets, the rise of the internet as a source of information for potential holidaymakers and the downturn in popularity of the far-flung holidays the holiday programme was presenting, in favour of tried and tested European package holidays and the 'staycations' that were widely touted in the British tourist industry as a product of the recession (see Walker, 2009). Indeed, as noted above, the disappearance of the BBC's longest-running holiday programme, *Holiday*, coincides almost exactly with the explosion of landscape programmes discussed at the start of this chapter; arguably, the soaring shots of Lake Windermere in *A Picture of Britain* or the presentation of the British coastline in *Coast* visualised alternative holiday destinations for many in the UK and thus inadvertently took over the holiday programme's raison d'être. Whilst BBC1 and ITV no longer produce high-profile, primetime, 'flagship' holiday programmes at the time of this writing, the holiday or travel programme *does* still exist in a number of guises. There is, of course, a dedicated Travel Channel placed deep in the EPG, with predominantly American programming and wall-to-wall episodes of *Airport 24/7: Miami* (2012–), and on BBC World News, a show called *Fast Track* has quietly been thriving since 2010, rebranded as *The Travel Show* at the beginning of 2014 with a repeated broadcast on Friday mornings on BBC2. Channel 5 also revived the primetime holiday programme in 2012 with their *Holiday: Heaven on Earth*, fronted by Emma Wilson, the daughter of ex-*Holiday* presenter Anne Robinson. In addition to these programmes, there have also been a number of reality/holiday hybrids such as Channel 4's *Holiday Hijack* in 2011 and RDF's *Holiday Showdown*, produced for ITV between 2003 and 2009.

If we take the BBC holiday programmes as an example, there have been several key aesthetic shifts in the genre since its inception, particularly since the turn of the century. Firstly, the studio base of *Holiday*, from which a series of BBC stalwarts presented their and others' roving reports on holiday destinations, was phased out of this show, to be replaced in the mid-2000s by a series of celebrities presenting from the locations they were covering in their reports. This more firmly situated the celebrity presenters within the 'imagined elsewhere' of the holiday programme. As David Dunn has noted, there was also a distinct turn away from featuring the opinions and activities of holidaymakers to a more singular focus on the presenting celebrity; in 2005 Dunn proposed that the holidaymaker's 'response to place is being replaced by an increasing foregrounding of the performance and celebrity of the presenter which, in turn, throws the tourist destination into a background of soft focus' (2005: 155). To a large extent, the presenter has always been central to the holiday programme, acting as a diegetic stand-in for the tourist; Dunn argues that these programmes offer 'surrogate sightseeing', in which the presenters 'are not tourists, but they have assumed touristic roles which inform their, and their camera's, narratives and performances' (ibid.: 158). This idea of 'surrogate sightseeing' can be related back to Lefebvre's 'enraptured character' who frames and defines 'intentional landscape'; this is seen, for example, in a report from the first episode of *Holiday 89* (tx. 5/1/89) featuring the presenter Kathy Tayler in situ enjoying a series of views in Dalaman, Turkey. As Tayler opens a piece to camera following a series of landscape shots, she exclaims 'Manzara. That's my first Turkish word and it means "panorama"'. Here she presents herself as an enraptured spectator, offering a performance of what John Urry defines as the 'romantic tourist gaze' (1990). Urry defined two separate versions of the tourist gaze, arguing that the romantic gaze is a gaze which is private, personal, suggesting a semi-spiritual relationship with the object (or landscape) of the gaze, whilst in the collective tourist gaze, other people are necessary to give atmosphere to the experience of place, and participate in a shared process of visual consumption. Whilst the presenter is seen 'revelling' in a moment of the romantic gaze in this clip, the presentation of such a gaze on television, revealed by Tayler's address to camera, is actually always an inclusive, collective gaze (in that it is broadcast to a collective audience).

The other key aesthetic and narrative shift in the holiday programme from its beginnings to the present day is actually a shift of redefinition, where the holiday programme becomes about *travelling* rather than holidaying. To some extent, the distinction between these two terms is defined by age and class: the old and the working class go on holiday, the affluent young, and the middle classes, go travelling. Staying with the BBC, the shift from *Holiday* to *The Travel Show* is also partly about a shift in the television landscape, a shift from the national address of a holiday programme made for primetime in a three-channel broadcasting era to the global address of a *travel* programme made for a 24-hour, globally broadcast news channel. As Morley and Robins have argued, televisual geographies are 'becoming detached from the symbolic spaces of national culture, and are realigned on the basis of more "universal" principles of international consumer culture' (1995: 11). The fact that the BBC's *Fast Track/Travel Show* addresses a global traveller who is just as likely to be watching from a hotel room in Dubai or his or her home in Copenhagen, rather than a UK home viewer shopping for holidays in the middle of winter, means that to a large extent the balance of reportage and the 'infomercial' aesthetic of earlier holiday programmes has been switched. Certainly, less emphasis is placed on showing an aesthetically pleasing landscape which might entice viewers to a particular location and more time is given to revealing socio-cultural histories of particular places.

When landscape is presented in a extended way in *The Travel Show*, there is a certain self-consciousness about the mediated nature of such a view; for example, in an episode broadcast in May 2014, an item on James Bond-themed luxury touring holiday in Scotland opened with a brief montage of aerial landscape shots. Here, a series of shots of an Aston Martin travelling through the landscape in *Skyfall* (Sam Mendes, 2012) were intercut with another Aston Martin travelling through the same scenery, shot for the *Travel Show*. Following the presenter's voiceover telling us 'And, for a price, Robert and his company will lend you their luxury Aston Martin, 007's car of choice, for a trip through this stunning landscape', we also get a series of point-of-view shots through the windscreen of the aforementioned expensive car. Here, the windscreen, one of the framing devices of the tourist gaze as defined by Urry (1990), acts as a screen, mimicking the cinema screen and immersing visitors in the intentional landscape of the big budget action film. In this sequence we see both a mediation of the imagined elsewhere, and the possibility of bringing the already-mediated imagined elsewhere into lived experience, albeit for a very high price. Whilst David Crouch, Rhona Jackson and Felix Thompson point towards the 'utopian unboundedness' of the tourist imagination (2005b), what *The Travel Show* actually shows us here is that the tourist imagination is often less about a utopian unboundedness and more about the pursuit of a mediated (or 'bounded') view associated with particular (classed) taste codes. We travel to search for a recognisable, pre-mediated view, in this sense, to experience views 'in the flesh' which have already been witnessed via film and/or television: this segment of *The Travel Show* visualises this quest for the viewer.

Finally, the reality/holiday show hybrids which became popular in the 2000s are also very interesting in their presentation of landscape. These include *Holiday Showdown* (RDF for ITV, 2003–9), in which two families are 'forced' to join each other's family on a holiday of their choosing, and *Holiday Hijack* (Betty TV for Channel 4, 2011), in which affluent British tourists are 'taken' out of their luxury hotels by a family from the local community who reveal the realities of life behind the glossy travel images that have enticed them there. To begin with *Holiday Showdown*, here we see what has become quite a standard reality television narrative of class conflict that might be understood as an encounter between holidayers and travellers, or between people invested in an unbounded utopian or hedonistic experience, and people invested in a search for a perfect view. The programme is made by RDF Media, which also produce *Wife Swap* (RDF for Channel 4, 2003–9), amongst many other reality programmes, and usually casts one (mainly working-class) family who enjoy package travel, the 'home comforts' of a resort with plentiful British food and lots of booze, and relaxed fun on hot beaches, and one set of 'cultured' middle-class 'travellers' who enjoy either further flung, more exotic holidays in which they engage more fully with local cultures and places defined as being of interest to sightseers, or explore the wilder reaches of the UK. It is immediately apparent when watching this programme that one of the ways in which the producers of *Holiday Showdown* articulate the class conflict at the heart of each episode is through the presentation of landscape. For example, in an episode in which the Wilson family (initially holidaying in Marmaris in Turkey) join the Bromhall family on a trek through the rainforest in Costa Rica (tx. 18/4/07), the mum of the Wilson family, Heather, is reluctant to be seen 'taking in the view' on her first morning in Costa Rica, following the standard montage of landscape shots over the top of the rainforest. What is important here is that an 'appreciation' of the beauty of landscape, the taking of pleasure in 'sightseeing', is implicitly understood as a classed activity by the show's participants. Heather's daughter expresses the sense of 'privilege' in being given access to a view which is normally beyond her reach; Heather then both acknowledges the beauty of the landscape, but is quick to point out that 'I'd get bored quite quickly looking at trees and vegetation'. Later in the episode when Heather is dismissive about a trip to see a volcano, Clive, the father of the Bromhall family, sneers 'She's trying to make her children ignorant. They might as well be sitting in the toilet at home for the amount of experience they're getting.' Here then, the producers of *Holiday Showdown* both offer the hesitations traditional to the holiday programme, in which spectacular, iconic landscapes are presented by a camera offering a combined romantic and collective tourist gaze, and they also select footage of their subjects and dialogue which presents the visual pleasures of landscape as part of an inherently classed set of taste cultures.

Holiday Hijack, on the other hand, is an interesting programme about the impact of tourism on local communities, and occasionally offers a more complex view of landscape. In episode four for example (tx. 21/8/11), in which a family from Essex is transported from a luxury holiday in a Kenyan reserve to the huts of a Maasai tribe on the Maasai Mara, we initially see humour and pathos in the family's inability to adapt to the landscape as experienced as *terrain*. Here, the dramatic conflict of the reality/holiday programme hybrid is about tourists struggling to *be* in the land; they cannot walk the distance from their hotel to the Maasai family's home weighed down by their copious luggage, and they struggle to undertake the everyday tasks of the Maasai people within this unfamiliar terrain. Ben, the head of the Maasai family they are staying with, is articulate about the complexities of the impact of the tourist industry on their lives: tourism has displaced them and draws heavily on their scarce natural resources, particularly water, but also brings in much needed revenue to the Maasai people, meaning they need to have control and agency in the local tourist trade. Through their encounter with Ben and his tribe, the Brookes family develop a different relationship to the Kenyan landscape, which is visualised by the programme makers in a key scene. This sequence, which offers the most 'iconic' shots of the wide 'empty' plains of the Kenyan Maasai Mara of the whole episode, images of what Marcella Doye would call a 'landscape stereotype' (2005: 24) figures the landscape as a backdrop to a discussion about the complex relationship between tourists/sightseers and the Maasai people. Of course, 'empty' is used in heavy inverted commas here; 'empty landscape' is what white colonials reported on seeing on their arrival in Kenya at the end of the nineteenth century (as discussed in Chapter 3), and to a large extent sweeping vistas of the Kenyan plain present this 'imagined' emptiness. Here the presentation of spectacular landscape, and the brief hesitation in the image to admire the 'intentional landscape' on offer, are complicated by and juxtaposed with the fact that the figures on and off screen in this sequence discuss the economy of an idealised view: what is lost by indigenous families in the attempt to 'preserve' a perfect view for tourists. Here, somewhat awkwardly perhaps, Colin McArthur's riposte to Colin MacCabe in the '*Days of Hope* debate' about radical television realism in *Screen* in the mid-1970s is brought to mind (McArthur, 1975/6). Briefly, McArthur proposed that television was capable of complexity and contradiction in its presentation of history: he focused on a moment from the radical historical drama *Days of Hope* (BBC1, 1975) in which the urbane mine owner tells three Durham miners that their assent will be won through non violent social democracy whilst on screen soldiers brought in to quell the strikers are in training, stabbing dummies with bayonets. McArthur argued that we must attend to what is shown as well as what is said when we evaluate a programme's 'message', or its success in presenting complex ideas or histories. Whilst *Holiday Hijack* is, of course, an entirely different kind of programming to this radical historical drama, in this sequence we see the contradiction between what is shown on screen (here the imagined elsewhere of Kenya as viewed by the tourist-eye of the camera) and what is being discussed in on-screen and voiceover dialogue. Here, the view shows us a spectacular landscape which is visually pleasurable and appealing to both the tourists on screen and the potential tourists watching at home; however, the voiceover narration and the dialogue of the participants express the 'Catch-22' situation (as defined by Morgan, the daughter of the family, on screen) of the tourist's search for a perfect view. We therefore witness television's potential to engage with the political economy of a view, and of place represented as landscape.

This latter analysis of the holiday programme extends beyond the exploration of the 'easy pleasures' in landscape discussed in relation to the first set of programmes discussed in this chapter. Arguably, in paying attention to the visual pleasures of British landscape programmes such as *Coast* and *A Picture of Britain*, more difficult questions about *territory* and the changing face of plural national identities discussed by Thompson (2010b) are left hanging. Whilst a 'spectacular view' on television is designed to appeal in a number of ways to a contemplative viewer or a potential tourist watching beautiful images in spectacular clarity, the question of what this contemplation might distract us from thinking about remains. In some ways this leads back to the 'spectacle as distraction' arguments that were explored in the introduction of this book: that looking or gazing at spectacular television might produce an uncritical viewer who is disengaged from the issues of socio-political importance explored, inferred by or ignored in a programme. Here the viewer is not watching television in a distracted way – according to Ellis' schema (1982) – but rather is absorbed in the programming in front of them and thus distracted instead from critically engaging with key issues about national identity, the global environmental emergency and so on. However, on the other hand, appeals to the viewers through moments of heightened visual pleasure on television can also produce a critical engagement, or, in the case of the television landscape, a sense of what is at stake in relation to environmental security: put plainly, if we are presented with the beauty of the natural world around us, we are more likely to want to protect that particular landscape for future generations.

Beyond this question of the push and pull of distraction and engagement that takes place around the presentation of the spectacular landscape, there is also the question of *pleasure*, of course. Spectacular landscapes punctuate contemporary television: from the lush green coastlines and vast deserts of *Game of Thrones* (HBO, 2011–), to the crane shots of the pretty English countryside that frame the action of *The Great British Bake Off* (BBC1, 2010–), landscape is carefully constructed to enhance the visual appeal of each programme. In these programmes and countless others, the presentation of landscape is fundamental to the visual pleasure provided by diverse genres of television programming. Whilst on the one hand the viewer watches these programmes in order for narrative enigmas to be solved ('Who will rule Westeros?'; 'Who will bake the best cake and win the competition?'), on the other, the appreciation of landscape is a critical secondary pleasure of these and other programmes. This chapter has thus explored intentional landscape as a key visual pleasure of contemporary television, arguing that the spectacular montage (a sequence in which the viewer is invited to 'just look') is key to television's presentation of spectacle, in order to conclude that these are pleasures that extend beyond the landscape documentary or holiday programme, and that permeate television programming more broadly.

Part III

Spectacular Bodies and (Tele)visual Pleasure

6

Fascinating bodies: Looking inside television's somatic spectacle

Across television genres, the human body has been presented as the object of an intense and scrutinising gaze. This chapter considers, for example, the rush of a CGI roller-coaster ride through the human digestive system, the televisual allure of fascinoma – those rare or unusual medical cases we are invited to gawp at on screen – and the limits of viewing pleasure when confronted with images of death and abjection on the small screen. Chapter 7 then turns to a consideration of the erotics of television and the powerful spectacle that desiring and desirable bodies provide. It is proposed here then that just as Linda Williams was able to identify the 'body genres' of cinema (the horror film, melodrama and pornography) in her exploration of the 'spectacle of a body caught in the grip of sensation or emotion' (1991: 4), so it is possible to define the body genres of television. In the next chapter, the dramas of sex discussed conform precisely to Williams' description of 'moved and moving' (ibid.) bodies caught in the grip of intense sensation and/or emotion; these dramas often inspire a bodily or kinaesthetic reaction in their viewers.[1] In this chapter, however, emphasis is placed on the body itself as an object of the gaze – the body, in fact, as an endlessly explorable entity which fixates and fascinates. Here an exploration of the televisual manifestations of what José van Dijck calls the 'transparent body' (2005) is offered, a series of complex mediations of the human body as infinitely visible and observable.

In John Caldwell's postscript to his exploration of the concept of televisuality, he takes the unusual turn of considering the notion of scopophilia in relation to television. Scopophilia, literally a love of looking, has been a central critical concept in the analysis of cinema, particularly since the publication of Laura Mulvey's 'Visual Pleasure and Narrative Cinema' in 1975 and its explorations of the gendering of looking and being looked at on film. Whilst Mulvey used the term 'scopophilia' to define a fascination with the objectified woman on screen, Caldwell, in his call for a return to an interest in the visual aspects of television, applies the term more literally. He critiques both the Freudian pathologising of the desire to see or gaze, and the fact that film studies evolved 'a not so subtle methodological loathing of the visual [...] sprung from speculations about scopophilia and the gaze' (1995: 343). Caldwell describes what he terms the 'scopophobia' of film theory, arguing that 'for high theory, the visual spectacle overwhelms and subjugates' (ibid.). He thus calls for a return to the visual in this work, arguing that 'the pleasures of the visual may be pathological and politically repressive as high theory states, but ignoring these pleasures is surely short sighted' (ibid.: 344). If Caldwell is correct in identifying this theoretical scopophobia across film, and subsequently television studies, then the following proposal is a radical one: that we explore the intense fascination with bodies on television, seeing the presentation of increasingly intimate spaces of the human body across televisual genres as a product of scopophilia, literally the 'love of looking', on screen. In this chapter, television's visual pleasures are often defined in relation to the pleasures of the visual investigation of the abject and the grotesque – the 'hard to take' and the 'difficult to see'. Nevertheless, just as we pull away from some of the intimate spectacles of the human body on screen, so are we also drawn in towards them; in this sense, they are properly spectacular, threatening to overwhelm us just as they hold us fixed in an intense gaze at the screen.

This work clearly comes out of and must engage with the broader 'corporeal turn' in the history of the twentieth and twenty-first centuries, and with that work which has become 'reflexively aware of the discursive parameters through which we access and interpret bodies' (Crozier, 2010a: 3); the body has become a significant site in history and in explorations of visual culture in the last half century, and it is thus no wonder that it has also become so present on our television screens. As Crozier proposes in his introduction to his cultural history of the body in the modern age, '[we] construct our identities – our inner selves – through relating to the "outside" world, to other bodies, internalizing other projections. To see other bodies is to see our self' (ibid.: 19). In the work on the place of anatomy in visual culture, the body has also been understood as a particularly significant site of the spectacular (Bull, 2012 and 2017; Cartwright, 1999; Stephens, 2012 and 2013; van Dijck, 2005; Waldby, 2000).

Attention to the body is not, of course, an entirely recent development for television studies. John Fiske proposed that the 'spectacular involves an exaggeration of the pleasure of looking […] Spectacle liberates from subjectivity. Its emphasis on excessive materiality foregrounds the body, not as a signifier of something else, but in its *presence*' (1987: 243). Fiske's work is critical in understanding the ways in which the body appears as the locus of the fascinated gaze on television and yet the impact of this work has, perhaps, been lesser than the impact of television criticism which developed notions of a more casual engagement with the television image. However, even such works normally associated with 'glance theory' as John Ellis' seminal *Visible Fictions* (first published in 1982) acknowledge that elements of television's visual grammar emphasise an intense gaze on *parts* of the human body on screen:

> Whereas the cinema close-up accentuates the difference between screen-figure and any attainable human figure by drastically increasing its size, the broadcast TV close-up produces a face that approximates to normal size. Instead of an effect of distance and unattainability, the TV close-up generates an equality and even intimacy.
>
> (Ellis, 1982: 131)

Ellis' 'even intimacy' here hints towards the intensity of intimate facial close-ups later explored by Karen Lury, who argues that the 'frequent proximity of the face and the emotions displayed means that the close-up on television is both sensational and, oddly, perhaps mundane' (2005: 30). Similarly, Milly Buonanno proposes that television's presentation of bodies in close-up is indicative of the intensity of our viewing of intimate spectacle across a range of programmes:

> [Television] offers up the faces and bodies of political and television personalities, protagonists in stories and events, ordinary people to the curious, empathetic or ruthless gaze of its viewers: faces and bodies that are full of symbolic indications to be read or scrutinized. We see tears welling up and flowing, we perceive unchecked reactions [...] sometimes we move closer to the screen to have a better look at a detail, if for no other reason than to experience the pleasure of staring unseen at a stranger without having to feel embarrassed by our impertinence.
>
> (2008: 40)

In Buonanno's terms, we hunt for clues of 'truth', motivation or sensation on the bodies of people in power, as well as on the bodies of characters in a drama, by staring into our television screens: we gaze 'ruthlessly' at the televisual body, and take pleasure in 'staring unseen' at such figures. Here then, television offers the appearance of intimate engagement with the televisual body, without having any responsibility for such engagement. As Buonanno presents it, television is the ultimate apparatus of the voyeur.

One area of television scholarship has taken up Caldwell's call to (re)turn to the visual and to consider the pleasures of looking on television, and examines both the 'excessive materiality' of Fiske's spectacular televisual body and dwells on Ellis', Lury's and Buonanno's quintessentially televisual close-up: the critical literature on reality television. Misha Kavka's work on the 'intimsphäre' of the genre and its bodily intimacy (2008: 50) and the broader work on the neo-liberal governmentality of reality television which produces an intense scrutiny of the bodies of the genre's participants, and which places those bodies and the bodies of the genre's viewers under intense pressure, has all figured the body on television as spectacular, or, more accurately, spectacularly presented as both unruly, abject and, then, ultimately, controllable or 'manageable'. For example, for Katherine Sender and Margaret Sullivan, US reality shows such as *The Biggest Loser* (NBC, 2004–) and *What Not to Wear* (TLC, 2003–13) instil in participants and audiences a willing acquiescence to surveillance and self-monitoring, and their audience research around these shows found their viewer-subjects expressing a combination of 'disgust and an unpleasant kind of identification' (2008: 576) in relation to the sensationalised presentation of a variety of 'unruly' bodies on screen. Similarly, Laurie Ouellette and James Hay

(2008) have argued, in relation to *The Biggest Loser* and *Honey We're Killing the Kids* (BBC USA, 2006-7), that through the sensationalised presentation of the body on reality television, the personal makeover is instrumentalised as a key responsibility of the good neo-liberal citizen. Other analysts of reality television have both emphasised the frequency with which we see bodies on screen, and they have discussed the visual grammar by which these bodies are presented as spectacle in reality television. For Diane Negra, for example, there has been, in recent years, a 'spectacular emergence' of the 'underfed, over exercised female body' (2009: 119) on television, whereas Beverley Skeggs and Helen Wood describe the 'cutting up' of bodies on reality television to create a dramatic focus on particular parts of the body as 'spectacles of shame'. Body parts, they argue, operate metonymically in this genre of programming to invite an intense focus on the body as a site of failure: 'In *You Are What You Eat* the close-up is extended from face to faeces as a symbolic representation of a badly managed life' (2012: 104). For all of these authors then, the body is caught in what Mark Andrejevic (2004) describes as a circuit of voyeurism and exhibitionism in which the viewer returns again and again to bodies that are both 'moved and moving' (L. Williams, 1991: 4), and presented as spectacle for an engrossed viewer who both identifies with, and is frequently either appalled and/or delighted by, the figures on screen.

The work on reality television is not the only area of television scholarship that has considered the spectacle of the body on television, though it is the most prolific site for its analysis. For example, the representation of the dead body in television crime and forensic drama has been a key point of debate within Television Studies (Brunsdon, 1998 and 2013; Creeber, 2001; Nunn and Biressi, 2003; Thornham, 2003, Sydney-Smith, 2007; Bull, 2012), and the critical writing on the medical drama is another important point of reference for this chapter, with its focus on medical documentaries and factual entertainment shows, particularly its definition of our relationship with the spectacular body on screen. Jason Jacobs suggests, for example, that the increasingly explicit and visceral focus of the medical drama is evidence of a 'popular fascination with decay, death, and the destruction of the body' and proposes that programmes such as *ER* (NBC, 1994-2009), *Casualty* (BBC1, 1986-) and *Chicago Hope* (Fox, 1994-2000) present 'a "morbid gaze" – the visualization of [...] horrible but routine body trauma'

(2003: 1). The morbid gaze of the medical or forensic drama is identified by Simon Brown and Stacey Abbott as a televisually specific form of body horror (2010a: 208) in which we are confronted with the fragility of the human body, and in which we 'revel in the spectacle of the grotesque body' (ibid.: 209). Whilst we see an image of the human body positioned as the object of a controlling or scrutinising medical/diagnostic gaze, we are also aware that our gaze may develop into 'a look of helpless compassion, horror, dread or morbid fascination' (Jacobs, 2003: 68). This fascination with the destruction of the human body in television drama identified by Jacobs in the early 2000s only seems to have increased in the last decade. Hospital dramas such as Sky One's *Critical* (2015–) and the period drama *The Knick* (Anonymous for Cinemax, 2014) have become increasingly gory: the pre-credit sequence of episode one of *The Knick*, in which a caesarian section in a public operating theatre in early twentieth-century New York goes spectacularly wrong, announces this increase in gore right from the start. In this case, Pete Boss' description of cinematic surgical body horror might be equally applied to medical dramas on television which 'are in many ways "about" this ruination of the physical subject; the fascination with this spectacle is not in any way a secondary consideration' (1986: 18).

Whilst this work on the body in television drama provides a significant critical context for this chapter, the focus here is on the spectacle of the 'real' human body on television as a less frequently discussed image: the bodies of real people, as well as those 'human' bodies constructed via a variety of programme-making techniques to demonstrate the inner workings of the biomedical body. This chapter takes three turns in order to examine this particular form of televisual spectacle. Firstly, it begins with a consideration of the 'how the human body works' documentary series, focusing in on several key examples of this enduring genre (e.g. *How Your Body Works*, BBC, 1958; *The Human Body*, BBC1, 1998; *Incredible Human Machine*, National Geographic, 2007; and *Inside the Human Body*, BBC1, 2011). This part of the chapter looks at how the body is framed by the televisual production of Foucault's medical gaze (discussed below), offering an analysis of the penetration of bodies on screen and the presentation of the body as a spectacularised 'fantastic landscape'. Secondly, the chapter will focus on documentaries of fascinoma – those shows that bring the 'extraordinary' body to television, – focusing on Channel 4's *Bodyshock*

(2003–) series and Channel 5's equivalent *Extraordinary People* (2003–), as well as the phenomenally successful, multi-platform *Embarrassing Bodies* (Channel 4, 2008–). In this section, the chapter will explore the notion of the televisual 'Freak Show', drawing on the history of this concept to explore the appeal of such programming and the economy of bodily visibility in the media. Finally, the chapter will explore the televisual presentation of the dead body. It will consider the presentation of actual death and the dead on television, and argue that whilst the corpse has been treated as a spectacularly dehumanised object (particularly in the television work of Gunther von Hagens), the treatment of death on television often moves away from the spectacular and towards a more ethical and caring treatment of the body and a humanistic representation that centres on the body as a being in time and place, rather than as the object of spectacle.

Gazing inside the body: Mysterious places and wild rides

All of the bodies in this chapter are fixed, to a greater or lesser extent, by the eye of the medical or clinical gaze, a concept introduced by Michel Foucault in *The Birth of the Clinic* in 1963 (2003) to describe the way in which the body had become conceived of as an object in need of medical interpretation, roughly from the beginning of the nineteenth century onwards. For Foucault, power is embodied in and comes with the day-to-day practices of looking at bodies associated with the work of doctors in the hospital or clinic, a look that he termed the 'clinical' or 'medical gaze'. Subsequently, and more broadly, his notion of the medical gaze has come to be understood as a term which describes the dehumanising medical separation of the patient's body from the patient's person or identity in clinical practice (see van Dijck, 2005: 11). In this work, Foucault provides us with a history of medical practices in the nineteenth and twentieth centuries in which the body is increasingly drawn into the realm of the visible and the readable, the understandable. Moving on and out from Foucault's history, the biomedical human body documentary on television forms part of a continuum of the spectacularisation of the body, a continuum which stretches from the public operating theatre of the nineteenth century to the human biology documentary series, the fascinoma 'shock doc' and the

TV autopsies of the twenty-first century. In the nineteenth century, public operations were common practice. Rooms for surgical procedures were called operating theatres because they literally were theatres, built in a gallery style for public observation. In the early nineteenth century, operations in institutions like St Thomas's Hospital in London were advertised in newspapers and surgeons might get a round of applause at the end of the procedure from the paying public; it was not unknown for an operation to be cancelled because public demand was so high that a larger theatre had to be found. Patients put up with having an audience at their time of distress (and threat to their life) because they received medical treatment from some of the best surgeons in the land, surgeons they otherwise could not afford (wealthy patients of the surgeons would have been operated on, by choice, at home). There is not much of a leap in creative thinking to be made between these medical spectacles, reimagined in the opening scene of *The Knick* as discussed above, and, for example, the 'media circus' surrounding the little girl undergoing corrective surgery in Peru who is the subject of the *Bodyshock* episode 'Curse of the Mermaid' (31/1/06), discussed below, or the seemingly endless parade of people willing to share their 'embarrassing bodies' on television for the chance to access specialist help.

Television is thus part of the ongoing process whereby that which was 'fundamentally invisible is suddenly offered to the brightness of the gaze, in a movement of appearance so simple, so immediate that it seems to be the natural consequence of a more highly developed experience' (Foucault, 2003: 241). Foucault speaks here of a development in medical knowledge (and, implicitly, medical imaging) which has fundamentally changed our relationship to the body and his proposal resonates with the fact that media technologies have developed alongside medical imaging technologies, often showcasing these new developments or, rather, new and more intimate views of the bodies in question. As van Dijck argues, 'medical and media technologies converge in their production of visual spectacle – displaying the inside of a human body' (2005: 10). Lisa Cartwright's work on medical visual culture in early cinema also makes this case:

> Cinematic apparatus can be considered as a cultural technology for the discipline and management of the human body, and [...] the long history of bodily analysis and surveillance in medicine

and science is critically tied to the history of the development of the cinema as a popular cultural institution and a technological apparatus.

(1995: 3)

Cartwright's important work has thus been to uncover a

history of the cinematic techniques that science has used to control, discipline, and construct the human body as a technological network of dynamic systems and forces […] [seeking] to demonstrate that this history is complexly interwoven with other areas of visual culture.

(ibid.: 4)

Whilst Cartwright's focus is solely on film, other works have broadened this analysis of the convergence of medical imaging and the media. Catherine Waldby (2000), for example, offers an account of the 'Visible Human Project', an early website produced by the National Library of Medicine in the US in 1994 in an effort to create a detailed data set of cross-sectional photographs of the human body, for which the body of executed murderer Joseph Paul Jernigan (and later other people) was cut into thin slices that were then photographed and digitised. José van Dijck's work (2005), on the other hand, offers a wide-ranging exploration of the 'transparent body' in the media and visual culture, exploring medical documentaries in film and television, the public anatomy exhibitions of Gunther von Hagens, educational websites and CD-ROMs, alongside the cultural history of key medical imaging techniques (the X-Ray, endoscopy, 4D ultrasound, etc.); this work, and particularly van Dijck's discussion of the surgical documentary on Dutch television, is central to the exploration of the spectacular televisual body which follows.

On British television, programmes *about* the body began in the 1930s, though it wasn't until the late 1950s that medical imaging played a major part in this programming. Initially, programmes such as *Body-Line* (BBC, 1937) and *The Limits of Human Endurance* (BBC, 1952) focused on showing what the body could do, what it could 'take'. *Body-Line*, for example, offered a series of physical demonstrations in five episodes: in one episode Margaret Morris (of the Margaret Morris Movement, a form of dance focusing on the breath and posture) brought in students to demonstrate her techniques to control the healthy body – other episodes featured physical education demonstrations from representatives of Men's Clubs, Boy's Clubs and keep fit classes for women. It was, however, the advent of schools programming which brought about the first human biology documentary series that was to incorporate microscopy, X-ray imaging and other body imaging techniques to show its viewers the inner workings of the body.[2] *How Your Body Works*, first broadcast by the BBC in the autumn of 1958[3] and presented by Professor W. S. Bullough (Professor of Zoology at Birkbeck College, University of London) ran for ten weeks and provided a series of illustrated lectures for secondary school children. Whilst the amiable Professor Bullough delivered the majority of the information about human biology speaking straight to camera in a lecture style, the programme was notable for the ways in which it integrated demonstration and visualisation into each episode. For example, in an episode called 'Food and Why You Eat It' (tx. 15/10/58), Bullough narrates a film in which he is shown swallowing a barium meal in front of an enormous X-ray machine. Following this sequence, shadowy shots of barium salts passing round the digestive system are screened whilst Bullough explains what the almost imperceptible images we are seeing are. Here the body is both brought 'to light' by the broadcasting of X-ray images, offering the viewer an intimate image of human biology, but it also remains a mysterious place in need of careful navigation by an expert 'guide'.[4] In his history of the clinic, Michel Foucault argues that 'clinical experience sees a new space opening up before it: the tangible space of the body, which at the same time is that opaque mass in which secrets, invisible lesions, and the very mystery of origins lie hidden' (2003: 150). Here, in these early programmes, is a very clear illustration of Foucault's point, as Bullough guides us through this 'opaque mass'. In another episode, 'The Heart and Blood Vessels' (tx. 4/11/58), increasingly magnified images, blurry and indistinct, of blood travelling round the veins and capillaries of a frog, are again explained by Bullough's narration, on and off camera, in the studio: he uses a pointer to interact with the image and navigate the extreme close-up on screen for the viewers, guiding us around features of the pulsing blood system as if they were points on a map. This schools programming thus again presents the body as a complex entity to be navigated by an on-screen expert. Furthermore, it is significant that animal bodies are frequently made to stand for the inaccessible human body in this programme; the frog's veins, a sheep's heart which is dissected on screen in the same episode, the mouse's ovaries presented in the later science documentary for a more general adult audience, *Life Before Birth* (BBC, 1960) all suggest that the human body continues to remain partly inaccessible, unrepresentable and beyond the reach of the television camera in this early period. By looking at this programme it is possible to see the beginnings of the promise of van Dijck's 'transparent body' (2005), but it would be several more decades before such 'transparency' would be achieved on television. Nevertheless, Granada produced a similar ten-part series for ITV schools programming called *The Living Body* in 1972,[5] and thus television is used in these programmes to enhance the science curriculum at schools, extending science education through a variety of body imaging techniques, demonstrations and the lectures of on-screen experts.

It is perhaps more surprising that throughout the 1960s, programming about the human body was more frequently located in the weekend factual entertainment schedules across both the main UK channels: on Saturday mornings (*Science on Saturday: Human Biology*, BBC, 1961), Saturday evenings (as part of the long-running science documentary series, *Horizon*, BBC2, 1964–), Sunday lunchtimes (*The Science of Man*, BBC, 1963; *Your Living Body*, ATV, 1969) and Sunday evenings (*The Wonder of Man*, ATV, 1960).[6] This scheduling is interesting as it firmly positions the human body documentary as 'family entertainment' with a broader address than the schools programme, situating this programming at the centre of a public service remit to inform, educate *and* entertain. Alongside the long-running Tuesday evening slot for science documentary in the UK,[7] this weekend programming sought to demystify the human body for a *general* audience, and subsequently to democratise knowledge of human biology via television. In 1974 Raymond Williams noted that 'visual demonstration of rare or complex material has markedly improved presentation of aspects of the physical sciences, of medicine' (1974: 54), almost certainly reflecting on exactly this cycle of programming. These documentaries, in their 'visual demonstration of rare or complex material' of biomedical science, thus looked forward to the populist address of the more recent documentaries to which we now turn.

Like the schools programming described above, a series of television documentaries from the late 1990s onwards, including *Body Story*

(Channel 4, 1998), *The Human Body, Incredible Human Machine, Miracles in the Womb* (Channel 4, 2007) and *Inside the Human Body,* used a wide variety of techniques to visualise the human body, drawing on new developments in medical imaging (4D ultrasound, microscopic 'pill' cameras, infrared imaging, etc.) and programme making (particularly in the field of animatronics and computer generated imagery) to provide a greatly more intimate and spectacular image of the detail of each system of the human body.[8] This programming is marked by its visual heterogeneity; it is truly televisual, in John Caldwell's (1995) understanding of the term, meaning that it is stylistically excessive and 'exhibitionist' in its aesthetic. It draws on this range of new imaging technologies both to tell the story of human biology more clearly, but also to 'show off' the modes of production by which this story is told, providing the viewer with a dizzying array of medical imaging and televisual styles. Whereas the body in *How Your Body Works* (BBC, 1958) needed to be narrated in order to help us understand the blurry and indistinct images on screen, in these more recent documentaries, the viewer is thrown from one spectacular image of the internal organs to the next, still needing to be 'guided through' the body on screen, but here in order to re-orient from each subsequent 'view' of the body. In each of these programmes, an opening montage of all of these different visualisations of the internal organs means that views of the inside of the body initially disorientate and disturb; here, the body is presented as a foreign terrain, a pink, pulsating, squelching, alien landscape shown from a series of different angles and in different ways. This reading thus aligns with Karen Lury's suggestion that such programming is filled with 'awesome or magical imagery' and her proposal that the 'wonder of the natural world or the complexity of scientific findings are matched and envisioned by the wonderful abilities of the camera operators, art directors and their visual technologies' (2005: 35). Foucault's 'opaque mass' of the mysteries of the human body is thus visualised in this programming as a fantastic landscape that can be traversed in visually spectacular ways.

To offer some examples of this trope, *The Human Body,* presented by Professor Robert Winston, opens, after a brief title sequence, with interior shots of the human body which are cave-like and mysterious, pink and shadowy chambers (actually parts of the respiratory system) which are referred to by Winston's voiceover as a 'unique dwelling place' he has been exploring for the last two years. Following this sequence, we see a pilot flying over the misty tops of a forest and then Winston himself in a variety of recognisable international places (the rainforest, the pyramids, the skyscrapers of New York), underscoring the idea that the body, too, is a *place* that can be navigated towards and around. The accompanying voiceover states, 'To come with me you'll have to cross the globe from Australia to Africa to America. You'll have to journey into space', at which point we cut from a shot of a space walk to zoom through a body's interior; the voiceover continues 'and into a place as mysterious but much closer to home'. As Winston says 'We've developed new techniques to help you get there', the camera zooms right into the skin of an optician who is inspecting his eyes on screen; the shot is then dissolved into a highly magnified image of the skin itself looking like a parched lunar landscape. As he announces that there are 'new cameras to show you the way', we see him swallowing an endoscope and then the subsequent image from inside his oesophagus. Throughout this entire sequence then, and in much of the following series, the body is a spectacular, navigable landscape. Motion is constant; we zoom, rush and fly around the body as if propelled by an aircraft or riding a vertiginous roller coaster, and the language of 'discovery' used throughout reflects this. We are 'taken on a journey', we 'dig deep' into the body and we are constantly invited to marvel at the sights on offer on this fantastic voyage. José van Dijck (2005) connects such images to the history and development of endoscopy, through which the body becomes a mediated virtual journey for medical practitioners, and discusses the incorporation of endoscopic images into the Dutch documentary series *Chirurgenwerk/Surgeon's Work* (Evangelische Omroep, 1990–). However, not all images of travel over a foreign (bodily) landscape are endoscopic: in *Incredible Human Machine*, for example, skin is described as 'a very different kind of landscape'. As an extremely magnified close-up begins to travel over the skin's surface, the timbre of narrator Andres Williams' voice drops: 'Magnified 600 times our outermost skin is nothing but dead cells, riddled with ridges and grooves and pocked with countless bumps and holes. Look closer still and we find hundreds and thousands of bacteria inhabit every square inch of us'. Here the camera moves over the body as an abject landscape, the voiceover inviting a sense of repulsion just as the extremely magnified image presents us with a place we have never been or seen before; the spectacular televisual body in this moment is an uncanny place, a place at once familiar and made strange, even threatening.

Whilst extreme magnification and endoscopic cinematography offer the viewer a spectacular body as a landscape which can be journeyed through, the incorporation of CGI into such narratives arguably amplifies this presentation. Van Dijck refers to this as 'virtual endoscopy' and argues that its use in the British docudrama *Body Story* enhances 'the illusion that virtual body inspections are painless and traceless [...] [They] appear to be pleasurable and even adventurous' (2005: 78). The opening of *Inside the Human Body* draws on the 'body as alien landscape' trope through exactly this kind of computer generated image of bodily travel. Over a tracking shot of a white, globular cell rolling over a spiky red surface, Michael Mosley's voiceover announces 'Hidden deep inside you is a wonderful, dynamic world'. Following shots of what looks like lava flow, accompanied by up-tempo, exciting music, he continues, 'Where vast forests of cells capture light. Where tiny movements trigger fierce electrical storms. And raging currents of blood feed your brain'. Here, the body is terrain, clean, smooth and pulsating red, but terrain which is presented as a space of danger, if not horror. Van Dijck traces such presentations of the body back to *The Fantastic Voyage* (1966), as the progenitor of those narratives in which characters shrink to miniature size to travel around the interior of the human body; indeed, this has also been an ongoing narrative trope on family/children's television. Episodes of *Dr Who* ('The Invisible Enemy' story arc, October 1977; 'Into the Dalek', tx. 30/8/14) and *The Simpsons* ('Treehouse of Horror XV', tx. 7/11/04) use this trope, for example, as does the children's animation *The Magic School Bus* (PBS, 1994–7), based on the books of the same name. Similarly, through the CGI 'fly through', these medical documentaries invite us to consider what it would look, sound and *feel* like to inhabit this 'mysterious terrain'. As van Dijck says, drawing on Bolter and Grusin's notion of 'remediation' (1998):

> Viewers who have grown up in the computer age [...] are no longer satisfied with representation. They want to be sucked in, soaked up by the object; they don't want to see the organ or molecule, they want to experience what it's like to be inside it [...] sight-seeing turns into site-seeing, as the distinctions between

the real, the represented, and the imaginative body fade into the
landscape of digital space.

(2005: 79)

Van Dijck proposes here a kinaesthetic experience of the human body in which these 'wild rides' around the body, recreated via CGI, offer us the thrill of virtual adventuring which we register in our own bodies as a kind of sensational spectacle. When in the *Inside the Human Body* episode 'First to Last', the closure of the hole in the heart of a newborn baby is presented in this way, we experience the kinaesthetic thrill of seeing and hearing a rush of red blood cells through the body, in a form of computer animation that mimics the motion of a fairground ride or water slide. In the later episode 'Hostile World' the progress of a fungal spore through the lung is also presented as a wild ride, accompanied by the language of sensational body horror.[9]

This spectacular, kinaesthetic audio-visual trope is described in Foucault's delineation of the *mobility* of the medical gaze which must

> travel along a path that had not so far been opened to it: vertically from the symptomatic surface to the tissual surface; in depth, plunging from the manifest to the hidden; and in both directions [...] The gaze plunges into the space that it has given itself the task of traversing.
>
> (2003: 166)

The language of travel and movement that Foucault uses to describe the medical gaze perfectly reflects the construction of the CGI sequences described above. The representation of body as landscape also draws on an earlier tradition of anatomical illustration that 'used illusionistic perspectives to endow the interior of the human body with a highly articulated depth, as much geological as geographical' (Sappol, 2010: 5); the use of computer generated imagery in particular enables this geological presentation of the body to be extended into the realm of the spectacular landscapes of science fiction in which alien worlds may be infinitely imagined and visualised.[10] In these programmes, then, to gaze is to travel, and the body as the object of this gaze is repeatedly represented as spectacular landscape.

Fig. 6.1 Body horror in *Incredible Human Machine* (National Geographic, 2007)

Whilst the computer generated images of the body as alien landscape are often rather beautiful (and in *Miracles in the Womb* they are romanticised further still by being accompanied by the specially composed poetry of Roger McGough and Brian Patten, each 'playing' a twin in utero), CGI is also used to present images of body horror in these documentaries. In *Inside the Human Body*, a baby skeleton crawls across the screen to demonstrate the mechanics of infant movement, accompanied by music more suited to a 'monster horror' movie. In fact, *Incredible Human Machine* presents a variety of horrific images, including a woman in the shower with mis-sized body parts to demonstrate the varying sensitivity of the body, and repeated images of flayed bodies, as in the shot of a woman standing at a mirror in the title sequence, pulling away the skin from her own chest (see Fig. 6.1). Whilst this image speaks metonymically of the documentary's desire to 'go deeper' into the human body and the fantasy of endless bodily permeability or 'transparency' (van Dijck, 2005) which this and other programmes promote, this is also a rather gratuitous shot; it is not instrumentalised elsewhere in the documentary to tell us more about the body's internal structure, but rather it is simply offered as an example of the 'excessive style' of spectacular televisuality, an invitation to the scopophilic pleasures of the abject and the grotesque.

Many of these programmes also take an important turn in relation to Foucault's notion that the medical gaze implies an implicit power relation between patient and medic, and it is a turn that relates precisely to the spectacle of intimacy which is endemic in television, as discussed by Kavka (2008) and Dovey (2000). That is, the viewers are repeatedly shown the medic-presenter offering his or her own body to the viewers as spectacle. This is a technique that was used in the 1958 schools programme *How Your Body Works*, where Professor Bullough's X-ray demonstrated the digestive system; in this sequence the doctor (or rather, zoologist) turns the camera on himself, inverting the medical gaze.[11] In the more contemporary programmes, this sense of bodily intimacy and penetration is exaggerated further: in *The Human Body*, for example, we see Robert Winston examine a cup of his own sperm, and allow a leech to suck his blood to demonstrate the parasitical relationship between mother and child in pregnancy. Michael Mosley, doctor-presenter of *Inside the Human Body*, has become known for his willingness to subject his own body to a variety of medical experiments, including infecting himself with tapeworms and then swallowing a 'pill-cam' to track their progress in *Michael Mosley: Infested! Living with Parasites* (BBC4, 2014). Winston and Gunther von Hagens[12] both penetrate their bodies with an endoscope for their respective programmes, and Dr Christian Jessen, the doctor-presenter of *Embarrassing Bodies*, discussed in the next section of this chapter, swallowed a pill-cam in a 2013 episode of the programme to demonstrate the digestive system and to underscore once again the fact that there are no boundaries to the intimacy of the views of the body which are offered in this genre of programming. Indeed, the argument presented implicitly in these sequences of the power inversion of the medical gaze is that if one can turn the medical gaze on one's self, then there is nowhere it can't or shouldn't go for the sake of informing, educating and entertaining the television viewer.

Freak shows, fascinomas and the medical economy of spectacularising the televisual body

Beyond the increasing intimacy of the human body documentary, what does it mean to present the medicalised body as spectacle on television?

A single turn of phrase has been repeatedly applied to programming that does this, a phrase which seems to encapsulate an uneasiness about looking closely at the televisual body: the 'television freak show'. In her analysis of the medical documentary on television, José van Dijck argues that

> The live freak show never really disappeared [in the twentieth century], but took on a new cloak; it evolved into the medical documentary, the appeal of which is based, to a large extent, on the convergence of medical and media techniques.
>
> (2002: 538)

Van Dijck thus proposes that 'Television is the road show or circus of the late twentieth century and media producers have taken the place of freak show managers' (ibid.: 552). Andrew Crisell (2006) also explores this idea in his analysis of the 2002 Channel 5 documentary *Dwarves in Showbiz*, which, he argues, exemplifies the ways in which disadvantaged people both exploit and are exploited by the entertainment industry. Several essays in Rosemarie Garland Thornton's collection *Freakery: Cultural Spectacles of the Extraordinary Body* (1996) also make the explicit connection between the freak shows of the nineteenth century and contemporary television (Stulman Dennett, 1996; Weinstock, 1996; Clark and Myser, 1996). For Crisell the 'shock documentary' is the epitome of the television 'freak show' in which a wealth of extraordinary bodies are presented, and which constantly threatens to topple its viewer into 'mere scopophilia' (2006: 81). He subsequently argues that 'if television is serious about documentary it should not exploit but *restrain* its own visuality – and that, in an age of fierce competition for audiences, is unthinkable' (ibid.). Here scopophilia (defined as a lesser pleasure of television by Crisell through his 'mere' prefix) is coupled with epistephilia,[13] and the desire to see threatens to overwhelm the desire to know. However, one might question exactly how television can or should '*restrain* its own visuality', or whether by placing such bodies on screen, the programme-makers reveal something about how these bodies appear within culture and society more broadly. By rendering these unusual bodies visible, or even spectacular, is the viewer forced to confront the ways in which they challenge societal norms, or the ways that they ask us to think about how we look at such bodies in everyday life? We might also ask what the spectacularisation of bodies in Channel 4's *Bodyshock* documentary strand, or Channel 5's *Extraordinary People* or the Discovery

Channel's *My Shocking Story* (2006–) reveals about the state of health care worldwide, or what it tells us about the economic imperatives of bringing such bodies 'to light'. The following analysis will show that whilst these programmes might be viewed as sensationalist and exploitative, they are also deeply self-reflexive about their position as twenty-first-century 'freak shows'.

One could not argue that these documentary strands provide the kind of public service claimed by *Embarrassing Bodies* (discussed below), alerting viewers to ailments and syndromes which they might unwittingly be suffering from: as both series are keen to stress, the conditions of the subject of each documentary are exceedingly rare. Typically, the voice-over of *Bodyshock* stresses this rarity in its opening preamble: 'Legs fused together like a tail, this child has one of the rarest birth abnormalities in the world' ('Curse of the Mermaid', 31/1/06); 'With just ten known cases in history, Manar is only the third baby [with two heads] to be born alive' ('Born with Two Heads', 20/2/06); 'Sometimes seen in Africa and Asia, this condition is extremely rare in the Western world – and it's destroying Warren's life' ('The Man with the Ten Stone Testicles', 24/6/13). As is evident here, the episode titles of these hugely popular documentaries[14] are intentionally sensationalist and pull the viewer towards a reading of the series as 'freak show', given that they draw on the presentational style of historical sideshow and travelling attractions with titles like 'The Living Mermaid'[15] or the 'Dog-Faced Boy'.[16] These opening voiceovers, and even the pre-programme announcements that precede each episode,[17] also draw on the tradition of fairground barkers, people loudly announcing the rarity of the fascinomas on display in their sideshows; thus these moments are the television equivalent of the call to 'roll up, roll up'.

These documentaries repeatedly and self-consciously call attention to the link between the freak show and the 'extraordinary body' narrative on television. Wesley Warren Jr., the subject of 'The Man with the Ten Stone Testicles', speaks to the camera in his introduction and announces 'This is a living and breathing freak show'; Michael, one of the subjects of *Bodyshock* episode 'Half Ton Man' (tx. 6/2/06) implores a news camera focusing on him being lifted into a specially reinforced ambulance 'Don't make it like a spectacle!… Please – it's not a circus act'. Indeed, in several of the *Extraordinary People* documentaries, the subjects are actually ex-freak show performers, and in these instances the equivalence between sideshow and TV show seems even plainer or more poignant. For example, the episode 'Octopus Man' (tx. 25/11/08) begins with the sound of 'spooky' tubular bells and a close up of 'Octoman' Rudy Santos' extra limbs, but not his face; here we are offered an image of Rudy as 'freak', as a depersonalised figure represented by his fascinoma alone. Leading into a voiceover description of Rudy's condition, the camera makes a slow zoom towards the signage of the sideshow where his body was initially presented to the public. This edit thus invites us to make a correlation between these two spectacular presentations of Santos' body. Following the title sequence, we are told that 'For 20 years Octoman was the star of the Philippine freak shows. He earned a fortune parading the extra limbs to thousands. He became a national celebrity. Some even compared him to Shiva, a mythical God.' The ensuing narrative, however, centres on 'finding' Rudy again, who is now living with his family in comparative poverty away from the sideshow, and introducing him to a doctor who will operate to remove his extra limbs, the remaining vestiges of a parasitic twin. We learn that Rudy is a kind and loving husband and father who is being made ill by his extra limbs, and we are invited to 'get to know him' as a fellow human being, rather than as the object of a sensationalist or spectacular gaze. The paradox of this documentary then is that whilst it, too, presents Rudy as fascinating corporeal spectacle for its viewers, frequently focusing in on his extra limbs and the small clump of his 'twin's' hair located on his stomach, at the same time it critiques and problematises this fascination.

A further paradox can also be found in the fact that whilst this episode, and other *Extraordinary People* episodes, such as 'Freak Show Family' (tx. 1/4/09), document the end of the freak show, and indeed the twenty-first-century freak show is often presented as evidence of the archaic or 'backward' nature of a non-Western country in these documentaries, they also might be seen as part of its perpetuation via the media – as van Dijck proposes of the surgical documentary more generally (2005) – Rudy's doctor comments on this when he says to the camera,

> Previously, many of these people would work as sideshow attractions in carnivals and circuses [...] The reason people like Rudy are no longer attractive is because they've lost their novelty, we can conjure up greater abnormalities on TV than in real life.

Similarly, 'Freak Show Family' documents the post-show lives of performers from a freak show in Indonesia, which was, according to the documentary, driven out of business when one of its key performers, Dede the Tree Man, appeared in the *My Shocking Story* documentary 'Half Man, Half Tree' (Discovery, tx. 9/1/08), rendering his extraordinary body too visible to the general public via television and the internet. Here media spectacle replaces the spectacle of the sideshow. In a poignant moment of introduction to Esih, the 'Melting Lady' who has an extreme facial disfigurement, Esih walks towards the camera at the start of 'Freak Show Family' as her voice is heard saying, 'I just don't want to be treated like an object'. Whilst her voiceover speaks of her desire for a life beyond the freak show, it also challenges the viewers to question their spectatorial position in relation to the image on screen, and it is important that she doesn't speak directly to camera. Her image on screen is thus positioned as spectacle: the voiceover implores us to look away.

The production of uncertainty or moral ambiguity in relation to viewing of the spectacularised body on television is frequently expressed in reviews of these programmes as an anxiety about the television 'freak show': Chris Shaw of the *Guardian* admits that 'one man's freak show is another man's portrayal of heroic triumph over medical adversity', but, he says, 'whatever you call it there's a lot of it about at the moment and there seems to be no shortage of willing subjects or eager viewer/voyeurs' (2006). Kate Bevan in the same newspaper berates herself for having watched:

> I rather despised myself last week for watching *The 34 Stone Teenager – Six Months On*, but even though I knew it was wrong, I couldn't help myself […] Why do we watch programmes like this? […] These are the modern equivalent of Georgian London's freak shows, when you could see giants and freaks in return for a few coins.
>
> (2007)

Zoe Williams' review of *Extraordinary People* also expresses this viewing shame: 'There is a constituency that finds pleasure in watching other people's sorrow and I have to admit I belong to it, though I hate that trait in myself' (2012). The self-loathing in these reviews might therefore be seen as evidence of the power of the television image to enthral and fascinate; here, disgust repeatedly lies not in viewing images of deformity and the 'abject' bodies on screen, but in the inability to tear oneself away from such spectacles.

As suggested in the introduction to this section, these documentaries are self-conscious about the media portrayal of the extraordinary bodies at their centre, and the economic imperatives behind rendering the body visible on a variety of media platforms. This self-consciousness is particularly present in the *Bodyshock* episode 'Curse of the Mermaid', in which Milagros Cerron, born with her legs fused together, is operated on by a top surgeon in Lima, Peru, under the full glare of a 'media circus'. As the introductory voiceover explains:

> Born to poor parents from a remote village in Peru's Andes mountains, her life is about to be turned into a melodramatic soap opera that will fascinate an entire nation. Ahead of her is a life-threatening operation to separate her legs. It will be coordinated by a charismatic and controversial plastic surgeon before an audience of millions on live TV.

It is made clear to the viewer throughout this documentary that, ironically, the programme makers disapprove of the extent to which Milagros' life is played out on camera in order for her to receive the treatment she needs. Her mother is shown seeing her baby for the first time via a television image, and the documentary also features scenes in which Dr Rubio arranges her christening live on television, with the local mayor as godfather, as a publicity stunt. Just before Milagros' surgery, in an operating theatre specially built to accommodate the live TV cameras and attendant members of the press, we see the body of the tiny girl remediated through a television screen within the shot, her body dwarfed by the media apparatus surrounding it. This image is a poignantly metonymical depiction of the depersonalisation of the medical-televisual gaze. Throughout the operation Milagros' doctors are shown to be worrying about the quality of the live feed of broadcast images from the theatre (and therefore their media portrayal as heroic figures), rather than expressing concern about the life-threatening procedure they are performing, and following the operation the Dean of Peru's School of Surgery is presented criticising this excessive media coverage. We thus see a circuit of publicity which enables Milagros' parents to access top-quality health care but which transforms their relationship with their daughter by *placing her on show*, and rendering her the spectacular object of a televisual gaze. At the end of the documentary, Sara, Milagros' mother, says 'Looking at her on TV you'd think she wasn't in any pain at all', over a remediated television image of Milagros who is just visible over Dr Rubio's shoulder. As the image slowly zooms into the pixelated body of the girl on screen, director Julia Harrington makes the point that this little body has become entirely mediated, entirely spectacularised, even for her parents. This televisual 'doubling', whereby the 'extraordinary body' documentary depicts previous, more sensational televisual depictions of the subject's body,[18] has thus become a repeated trope in this documentary strand as a way to self-consciously critique the gaze on the spectacular body. Time and again, episodes focus on the subjects of these documentaries being 'forced into' media appearances as a way to access otherwise unaffordable health care. The *Bodyshock* episode 'The Man with the Ten Stone Testicles' follows the rise and fall of Wesley Warren Jr. as a media celebrity as he tries to raise money to pay for exorbitant US health care or persuade an experienced plastic surgeon to perform pro bono surgery on him. The narrative thus shifts from its initial focus on Warren's testicles as spectacle to the nail-biting 'will he raise enough cash to persuade Dr Gelman to operate on him' story arc. Whilst Gelman refuses to allow Warren to auction his severed testicle sac to raise the money to pay him, we later see his 'heroic' doctors 'clinking' the testicles together and saying 'Cheers' on camera in a moment that speaks of the economy of the spectacular body on television. For his doctors, who found Warren's attempts to directly monetise his body distasteful, their appearance on television advertising their role as his heroic saviours is worth celebrating. As Michael Sappol writes of the nineteenth-century doctor: 'The spectacular case, the rare and freaky case, was particularly useful in dramatizing the itinerary of pathology in the body and advertising medicine's (and individual doctor's) profound knowledge of it' (2010: 4). In the end, nearly all of the episodes of *Bodyshock*, *Extraordinary People* and *My Shocking Story* become narratives of surgery; they become not just about the spectacle of the extraordinary body, but also about the spectacle of surgery[19] and the 'mastery' of modern medical science over these spectacular bodies.

These documentary strands thus often position the economic necessity of rendering one's body as televisual spectacle as a problem of developing nations, or places with less progressive attitudes to disability or without the provision of a national health care system; however, in the UK, *Embarrassing Bodies* also reveals a similar economy of bodily spectacle. Described by *Evening Standard* television critic Victor Lewis-Smith as 'admirable, unpalatable, fascinating and repulsive in roughly equal measures' (Wiseman, 2010), *Embarrassing Bodies* started life as *Embarrassing Illnesses*, a 30-minute factual entertainment programme in 2007 and has been broadcast every year since then, branching out into a variety of spinoff shows[20] and its live road-show version, *Embarrassing Bodies: Live from the Clinic* (Channel 4, 2011–). It is a huge ratings success, gaining roughly double Channel 4's usual audience figures for its 9–10 pm slot, and presents a core team of doctors and visiting specialists offering sympathy and help for patients who come to their clinic/studio with 'embarrassing' medical problems. This advice is crucially coupled with frank and intimate examinations where every part of the body is displayed in graphic detail, along with health advice and footage of corrective operations. The programme is accompanied by *Embarrassing Bodies Online*, the interactive element of this cross-platform 'project' and Channel 4's most frequently visited site;[21] the site incorporates information about key preventive health topics with networked video, games and other applications. Spikes in web traffic for the site around the broadcast of each episode (Wiseman, 2010) show that viewers 'multi-screen' during the programme, suggesting that television's corporeal spectacle is capable of crossing media sites, and that, in fact, the internet in this case acts as an extension of the extreme intimacy of bodily spectacle on television.[22] Unsurprisingly, the 'freak show' nomination has also been used by television reviewers to describe the appeal of this programme. Julia Raeside asks, for example, 'Are these shows really raising awareness of taboo maladies and emboldening a notoriously awkward British public to see their own GP? Or are we just Victorians copping a look through the bars of a freak show?' (2012).

The aesthetic of *Embarrassing Bodies* in many ways emphasises the presence of the surveillant, medical gaze; the studio, a white box in which a series of monitors display the bodies of its subjects, via webcams (in the case of *Embarrassing Bodies: Live from the Clinic*) or pre-filmed sequences in the studio, contrasts the clean, white, sterile, *controlled* medical environment with the smelly, sore, weeping, *uncontrolled* bodies on display. In bringing the body into this visual environment, the doctor-presenters begin a process of both spectacularisation and containment of the 'embarrassing' body. One question that is repeatedly, and understandably, asked of the programme, however, is why its subjects bring themselves into this environment and offer themselves as televisual spectacle? Raeside expresses this when she asks 'Who are these people? Where does the idea come from that only a TV expert can cure your ills?' and goes on to state 'We've got an NHS for pity's sake. Even an over-stretched GPs surgery is better than the full public humiliation on offer here' (ibid.). In Eva Wiseman's interview with a number of viewer-participants who attended the *Embarrassing Bodies* truck, which travels round the country acting as a mobile surgery-cum-studio, her respondents frequently intimated that they did not believe they would be helped with their worries about their bodies by the NHS:

> [One woman said] 'I love the show. And I love the doctors that make themselves so approachable. GPs, usually, are so busy they don't have time to listen' [...] Kelly Coulter, who's brought her 18 month old son to the truckstop to talk about a problem with his gums, says she'd 'absolutely get my breasts out on the show if I was guaranteed a boob job'.
>
> (2010)

So whilst *Embarrassing Bodies* is often held up as the prime example of Channel 4's public service television, educating viewers about a number of common health problems and reducing the stigma of talking about worries that they may otherwise hide, it also implicitly exposes some serious problems with a struggling national health service. Take, for example, a segment of *Embarrassing Bodies: Live from the Clinic* first broadcast in November 2013,[23] in which Dr Christian discusses the treatment for Xanthelasma (patches of thick yellow skin formed on the eyelids) with a viewer-participant via webcam. Whilst he tells her that a chemical skin peel may help her, he acknowledges:

> I think sometimes these sorts of treatments can be difficult to access on the NHS. It's all about priorities, and cosmetics, and funding and things. But certainly I think we can help you out with this and refer you to someone who could discuss various options, so a chemical peel or even laser removal would work well for you, so we'll do that if you like.

Throughout this series, the 'pay off' for subjecting one's body to the scrutinising gaze of both doctor and viewer in a dual 'medical gaze', and the pay off for spectacularising the body on television, is access to private health care which bypasses an NHS which is frequently seen as unable to cope with demand. The economy of corporeal spectacle explored in the 'extraordinary body' documentary (access to 'free' health care in exchange for the spectacularisation of the 'sick' body) is thus also present in this more domestic of medical entertainments.

The dead and the dying: The limits of corporeal spectacle on television

The spectacular bodies discussed thus far in this chapter have all been presented on television as evidence of the persistence of life. However, in the conclusion of this chapter, a turn is made to the dead and dying body. In 2014, the spectacle of death on television (and its absence) was particularly significant. ISIS/ISIL, a Sunni extremist, jihadist group based in Iraq and Syria, released a series of horrific propaganda videos aimed at Western countries in which two American journalists,[24] a French mountaineering guide[25] and three aid workers[26] were beheaded. These videos, rather more 'slickly' produced than previous videos released by terrorist organisations, attempted to harness the power of the media to terrorise the enemies of ISIS/ISIL by making a gruesome spectacle of death on television and the internet. The existence of these videos placed television news editors and organisations in a difficult position, needing to balance the requirement to represent this atrocity with a desire not to acquiesce to the spreading of images of terror. As Al Jazeera America stated, 'We have not and will not air any video or audio from the horrific tapes […] These videos are used for propaganda and we want no part of that. We can report the facts without showing the images' (Steinberg, 2014). There was a general acceptance from US and UK broadcasters that whilst these images would not be shown on television, they would be freely available online, despite attempts by government agencies to have them removed, and despite appeals from both the victims' families and a widespread Twitter campaign for people not to access them. Thus these powerful spectacles of death, unacceptable on television, would be viewable on a less carefully edited and heavily regulated medium.

Documenting such instances where the horror of the dead and dying body might be presented via television news, Dan O'Connor argues that there remains:

> a pervasive belief that suffering, dying bodies [retain] the power to traumatize those who [see] them [...] [and] in the general unwillingness to broadcast the death of a human body, a retention of the belief in respect for the corpse, a belief as old as human civilisation.
>
> (2010: 124–5)

This echoes André Bazin's work in which he compared the screening of a real death to 'the profanation of corpses and the desecration of tombs' (2003: 31).[27] Broadcasters thus are given the implicit responsibility to represent, or sometimes not represent, the dead body with care: to thus refuse to spectacularise the dead or dying body. Whilst news broadcasters regularly work through these issues, controversy has also surrounded the depiction of the dead body in other areas of programming. Channel 4 in the UK caused a minor 'media storm' in its representation of the dead in the arts documentary series *Vile Bodies* (1998), and particularly the work of photographer Sue Fox and her images taken in a Manchester morgue (see Treneman, 1998). Later, Ofcom censured Channel 4 for showing images of performance artist Zhu Yu, eating the corpse of a stillborn baby in the programme *Beijing Swings* (tx. 2/1/03), arguing that the broadcaster had shown 'a lack of respect for human dignity' (Deans, 2003).[28] Comparing the latter broadcast to the work of anatomist Gunther von Hagens, Ros Coward argues:

> It doesn't matter whether some alcoholic old tramp gave von Hagens permission to dissect him or an impoverished Chinese mother accepted a small fee to allow her dead baby to be

> cannibalized. That just makes it sadder. Broadcasters should know it is dehumanizing to show such acts to what could have been an indiscriminate audience. To have such a thing flashed up on a screen, or watched out of passing curiosity, undermines an unspoken, collective feeling that however insignificant a person has been, we regret death and treat it with respect.
>
> (2003)

Coward's anxiety here is thus linked to the specificities of television broadcast: if images of the dead and the violation of their bodies are integrated into the *flow* of television, does the casual viewer stand to be caught in the glare of an unexpected (and, for Coward, immoral) form of televisual spectacle?

For Elisabeth Bronfen, the image of the dead body is restorative, reassuring: images of death

> delight because we are confronted with death, yet it is the death of the other. We experience death by proxy [...] The aesthetic representation of death lets us repress our knowledge of the reality of death precisely because here death occurs *at* someone else's body and *as* an image.
>
> (1992: x)

Following such an argument, originally constructed in relation to images of the female corpse in fine art, an encounter with the dead via television would reassure the viewers that death is elsewhere, offering a kind of catharsis in which they are reminded of their own vitality. However, an alternative view might be to argue that the very intimate and often graphic images of the dead on television (in contrast to those paintings and sculptures Bronfen focuses on) offer a far more abject spectacle that fails to reassure in this way. For Julia Kristeva, confrontation with the corpse, and thus the materiality of death, is more disturbing, and is seen as the defining moment of abjection:

> corpses show me what I permanently thrust aside in order to live. These body fluids, this defilement [...] are what life withstands, hardly and with difficulty, on the part of death. There, I am at the border of my condition as a living being.
>
> (1982: 3)

The viewer sees the corpse, and experiences it in the (televisual) moment, in order to be confronted with the horror of death. As I have argued elsewhere (2006), the presentation of images of horror via television has a particular potential to upset the viewer: television disruptively brings such images into the intimate 'sanctity' of domestic space, and the flow of images on television renders it potentially uncanny in its ability to bring together the familiar (the domestic) and the strange (in this case, death).

It is interesting in relation to this discomfort in broadcasting death and the dying, and the fear of being confronted with the abject on screen, that, as Mark Lawson states, 'it was common in the first decades of the medium for critics of TV to suggest that prime-time executions would be the logical outcome of mass visual entertainment' (2014). Whilst these early fears about the medium leading, inevitably, to the broadcast of hangings were widespread, in another recent example of the coverage of conflict in Iraq, television news initially refrained from broadcasting precisely such a spectacle. In 2007, news coverage around the world of the execution of Saddam Hussein incorporated the syndicated images of the events leading up to the death of Hussein provided by Iraqi state television Al Iraqiya: however, all initial television coverage stopped at the point at which the noose was placed around Hussein's neck, and thus the spectacle of his death occurred off screen. It was therefore an absent spectacle of a brutal death, imagined in the minds of viewers, but not present on screen. However, in the days following the hanging, mobile phone footage taken by one of the guards at the execution emerged, showing much more graphic images of the hanging accompanied by the sounds of the dictator being jeered as he died. Once this unabridged footage was freely available and distributed widely online, television news programmes and websites began to incorporate excerpts, images and transcripts from it. 'Prime-time executions' are thus 'the logical outcome of mass visual entertainment' only in the fact that television's presentation of Hussein's death is pushed to further sensationalism by more freely available imagery online. Lawson made this point about the televised execution specifically in relation to a contemplation of the television autopsy, to which this chapter now turns.

No account of the spectacle of the 'medicalised' human body on television could fail to consider the work of Gunther von Hagens or what it tells us about the visual culture of the twentieth-first-century body. Von

Hagens, a German anatomist, has become widely known for his *Body Worlds* exhibitions, large-scale, touring public exhibitions in which a number of corpses, preserved using the plastination technique he himself developed, are exhibited in a variety of poses that demonstrate different aspects of the human anatomy but which also position these exhibits at the cusp of a kind of macabre sculptural art. Describing the exhibits included in the first UK exhibition in 2002, Mark Espiner writes

> Twenty six corpses are on display, including the flayed cadaver of a man holding his own skin like a raincoat, a man bestride a horse carrying his own brain in one hand and the centerpiece: an eight-months pregnant woman with her womb opened to reveal her foetus.
>
> (2002: 15)

These exhibitions have proved hugely popular worldwide[29] and continue to tour to the date of publication. Von Hagens stresses that *Body Worlds* is made using the bodies of volunteers and that it is primarily intended to educate the public about the human body (the exhibitions have a 'public health' message about taking care of the body tied into them) despite concerns about the propriety of presenting the dead body in this way. Whilst a number of theorists have considered these exhibitions and their role in broader histories of public spectacles of anatomy (see van Dijck, 2005; Stephens, 2012 and 2013, there has been no sustained consideration of von Hagens' television programming as an extension of his popularisation of human anatomy via the presentation of the dead body.

At the end of 2002, von Hagens was commissioned by Channel 4 to perform the first television autopsy, broadcast a few hours after he performed the autopsy in front of a small audience in a gallery in east London. This television event, a 'resurrection' of the public dissections described earlier in this chapter, attracted a huge amount of press interest, as well as threats of charges for von Hagens after the national Inspector of Anatomy, Jeremy Metters, ruled the autopsy illegal as von Hagens 'did not have a license to carry out postmortem examinations under the Autopsy Act, nor was the gallery licensed to hold them' (Gibbons, 2002a: 6). In the end, the broadcast went ahead without charge and 1.4 million viewers watched, 'forming one of [Channel 4's] best late night ratings of the year' (Gibbons, 2002b: 7).

Following this broadcast, von Hagens made four further programmes for Channel 4: the anatomy/autopsy series *Anatomy for Beginners* (2005), *Autopsy: Life and Death* (2006),and *Gunther's E.R.*, and the single documentary *Crucifixion* (2012), which couples the story of his construction of an anatomical exhibit based on Christ's crucifixion with his own struggle with his failing health. This programming has, in turn, spawned a further cycle of autopsy documentaries including the animal autopsy show *Inside Nature's Giants* (Windfall for Channel 4, 2009–) focusing on the spectacle of the internal organs of large animals; *Prehistoric Autopsy* (BBC2, 2012), described on its website as 'a journey into our evolutionary past, piecing together the bodies of our prehistoric family'; and *Autopsy: The Last Hours Of...* (Channel 5, 2014–), combining witness testimony with the interpretation of coroner's reports into the death of various celebrities.

Explorations of von Hagens' exhibitions have seen them as contemporary revisitations of the work of the anatomists who demonstrated their skills of dissection in the European theatres of anatomy in the eighteenth and nineteenth centuries (Jeffries, 2002; Stephens, 2012 and 2013), and certainly von Hagens has a similar penchant for theatrical spectacle, as is evident in his television documentaries. In the first moment of his series *Anatomy for Beginners*, for example, a carefully choreographed hour-long dissection to show the processes of human movement, von Hagens appears in his trademark fedora hat in front of a body which appears to be hanging by the head from a hoist, its face covered by a blank mask of plaster of Paris. The hat, the ascetic wire spectacles, the drawn face which resembles that of classic horror actor Lon Chaney Sr,[30] the darkness of the studio and key lighting behind the cadaver in the rear of the shot all refer knowingly to a history of the horror film; indeed we cannot watch this moment without this history in mind (see Fig. 6.2). The removal of the skin of this cadaver in a single piece following the title sequence is a stomach-turning performative flourish, rather than essential for the anatomical science of body mechanics discussed in the episode, and again demonstrates the skills of von Hagens as a 'show man'. Similarly, he opens the first episode of *Autopsy: Life and Death* with a soliloquy about the nature of our death, standing before a studio full of body parts, lit with blue lighting, and holding a plastinated skull up to his face in a nod to a moment of classic body horror from *Hamlet*. When he closes this episode with the line 'This,

Fig. 6.2 Gunther von Hagens with cadaver in *Anatomy for Beginners* (Channel 4, 2005)

indeed, is a quite unusual programme and some of you may have been frightened in the beginning, but fortunately enough, nobody was frightened to death', the viewer sees and hears his self-conscious performance of the purveyor of body horror, like the TV host of a late night horror movie marathon.

In contrast to these macabre openings, the studio in which the dissections take place is a pristine 'theatre of surgery' rather than the chaotic, Gothic workshops of Dr Frankenstein et al.; flanked by raked seating and a studio audience on three sides, the studio is filled with the corpse, a live nude model, plastinated bodies and body parts, all navigated by von Hagens and his co-presenter Professor John Lee (pathologist from Rotherham General Hospital and Hull-York Medical School). The presence of the studio audience as diegetic spectators focuses the viewer's attention on the experience of looking at the spectacle of the corpse at the centre of this space, as does the sparseness of the studio design. Just as the viewers are drawn to stare at the body on screen, so are they also drawn to examine the faces of the studio guests (in the latter two series, all potential donors and their families) for signs of their experience of spectatorship: some look rapt and engaged, whilst others look around them in discomfort, but their presence reminds us of the process of looking and edits to their faces throughout each dissection underscore this. As Elizabeth Stephens argues, 'the inclusion of the shocked reactions of the audience in *Anatomy for Beginners* [...] constitutes a deliberate construction of such events as both confronting and titillating' (2013: 128). Whereas once the bodies of criminals, post-execution, were the subjects of public dissections, the public autopsy as a form of punishment and humiliation beyond life (Richardson, 2010), here we are also drawn to wonder about the motivation behind offering the body as public, televisual spectacle in death. The studio audience is also important in understanding why the viewers might read these programmes as offering the spectacle of abjection, given each episode brings corpses, and the body parts of the dead, up close to the bodies of the living. The audience members squirming in their seats or gazing intently at the sight before them acts as our diegetic stand-in, a living body on screen which represents the viewer in their encounter with death and the horror defined by Kristeva as the horror of 'death infecting life' (1982: 4).

The starkest moment of the horror of the meeting of dead and live bodies is at the beginning of the final episode of *Autopsy: Life and Death*, 'Time' (tx. 6/2/06) where Anna, an 84-year-old woman, stands naked next to von Hagens. Looking away from Anna and towards the camera, he says 'Anna is 84. Her body has changed throughout her life. It's more fragile than [it] used to be.' The camera then zooms towards von Hagens' face as he asks why some of us will live to this age and others don't. Following this move, he walks over to a cadaver, announcing: 'Tonight I will try to answer these questions [about death] by showing you this 84-year-old woman [gesturing to the corpse] sliced in half'. In the rear of the shot, Anna's body stands, her faced turned away from the camera, her hunched back and papery, pale skin lit by key lighting from above. Here both these women's bodies, the living and the dead, are held together for an instant, but long enough for the viewer to experience the horror of abjection on Anna's behalf and to think about how she is feeling: Is she cold? Embarrassed by her nudity? Upset by the proximity of her body to the corpse of the woman her own age about to be dissected by von Hagens? This episode, and particularly this opening sequence, is painful to watch. The viewer experiences it kinaesthetically as well as emotionally. Sue Thornham argues in her discussion of the fictional autopsy that there is something personal in our response to the figure of the dead body, and specifically the dead woman, on screen. This comment resonates with my own discomfort in this moment, given that I identify strongly with the women on screen who are subject to the powerful, dehumanising medical gaze of the television camera and von Hagens himself. Thornham argues that 'For the cultural analyst who is also a female spectator, there is always something personal as well as professional at stake in these representations [of naked and dead bodies]: it matters who is "staring at bodies" and who speaks' (2003: 91). Here the body as televisual spectacle feels like cruel and dehumanising image.

Von Hagens claims that his work demystifies death and that this bringing together of the dead and the living, even in a sensationalist or spectacularising way, is an important public service. Cultural historians have indeed corroborated the fact that we live in an age where death is held at a distance, and that it has not always been thus. Ruth Richardson, for example, argues that 'the transition from life to death was much more familiar in every previous era than it is today', conceding that 'Of course we see deaths all the time on television and in film, but they are not *real* deaths' (2010: 93). Whilst Richardson might argue that even depicting *real* deaths on television holds the viewers at a distance from them, offering a mediated image of dying rather than real world (and real time) access to this experience, the desire to represent the processes of death and dying on television speaks to this cultural absence. Richardson's point about the absence of death is also made in Michele Aaron's exploration of death and the moving image. On the one hand, Aaron proposes the following:

> Death is everywhere and nowhere in contemporary Western culture. Corpses litter Hollywood film [...] [and] the recently recovered or slowly dying make our bookshelves groan. But the pain or smell of death, the banality of physical or undignified decline, the dull ache of mourning, are rarely seen.
>
> (2015: 1)

However, in her exploration of documentaries that depict death, particularly Allan King's *Dying at Grace* (TVOntario, 2003), Aaron argues that these works work hard to bring the experience of dying to light. Of the latter film she argues that:

> it offers an extraordinary commentary on the capacity of film to represent dying, to express it and share it, and to do so ethically,

> that is, with an eye, or better still, an ear to the experience of dying and to our (deprivileged) part in its lesson.
>
> (ibid.: 159)

To return to the human body documentary, it is telling that in both *The Human Body* and *Inside the Human Body,* the episode dealing with death is markedly stylistically different from the rest of the series. Narrative pace is slowed right down, the visual style shifts to that of the more familiar television documentary without the endoscopes and CGI whistles and bells of earlier episodes: here, our focus is drawn to the *person* rather than the body, and his or her experience of death. Each of these episodes works hard to convey the experience of time for the dying and for those who are left behind by them. The episode 'The End of Life' from *The Human Body,* for example, begins with a series of 'bullet time' shots in which everyday activities are suddenly 'frozen' and then tracked around. This self-consciously emphasises a Barthesian understanding of the photographic image as being inextricably linked to death or the idea that the photographic image always offers us the 'return of the dead' (Barthes, 1981: 4). For Barthes, the ability of photography to memorialise a person is related to what he identifies as the essence of the photograph: the that-has-been. This stylistic flourish, then, aims to point us towards the fact that the television image may be understood as an extension of Barthes' photograph, capturing the that-has-been in its representation of the dying man whose story follows. Accompanying these images, Robert Winston confirms this reading when he says, in voiceover,

> Often, our only experience [of death] is from films and television which can present it as a violent and painful event. We are reluctant to face up to our own mortality, to confront the truth that in the midst of life, we are in death.

The latter half of the episode subsequently follows the story of the death of Herbie Mowes, a German man in his sixties who is dying of cancer, and is living in Ireland with his partner Hannelore. Here, as in other television presentations of death, such as the documentary about assisted dying, *Terry Pratchett: Choosing to Die* (BBC2, tx. 13/6/11), the body of the dying man is not the focus; we see Herbie's body only in relation to his home, his family and carers, and his own frustrations with its failing, and no detailed visualisation of the biomedical body or his symptoms is offered. Strong objection was raised to the broadcasting of Herbie's death: for example, John Beyer, General Secretary of the National Viewers' and Listeners' Association argued

> Showing the moment of a man's death on television is totally inappropriate [...] Our organisation as a whole thinks that the filming of somebody dying is totally wrong. We believe that the moment of death is a very private matter for the individual concerned and their relatives [...] not for television viewers. Filming this time with a camera in a dying person's face, we feel, is totally inappropriate. And it is certainly not a moment to be shown on TV.
>
> (Rowe, 1998)

However, ultimately there was a wider sense that the death had been sensitively portrayed in a non-spectacular way (Aaronovich 1998), and that Winston's defence of the representation of death *as part of life* had been correct. There is a lyricism to this sequence of the episode, which includes extended footage of his friends coming to sing to him as he dies. Similarly, in the Pratchett documentary, repeated, elegiac shots of absence are offered to visually represent this sense of time passing and eloquently express a sense of loss: misty, grey exterior shots; shots of the night streets of Zurich as Pratchett visits people bound for a 'good death' at the Dignitas clinic. In actual fact these programmes show us the impossibility of representing the experience of death on television or elsewhere. Elisabeth Bronfen argues, via the work of Walter Benjamin, that death is ultimately unrepresentable; it is the 'one experience that always recedes from our knowledge' (1992: 80). Similarly Michele Aaron explores, via the work of Elaine Scarry (1985), 'the unshareability and inexpressibility of dying in culture' (2015: 99), though Aaron ultimately refutes this position when she says that 'Given the importance of the affective register to film – how it is a *moving image*, it emotes, is emotive – spectatorship could be thought of as symbolizing shareability itself' (ibid.: 155). Represented death must, however, offer an external and imagined view of death, often from the perspective of those who grieve, but not the experience of death itself. So, whilst this section began with the horror of death as spectacle, it ends by acknowledging that death is always just beyond us and that the confrontation with the materiality of the dead body via television can only be a stark reminder that we are forever positioned to never really *know* death, despite attempts by programme makers to make us intimately acquainted with it.

Von Hagens' television programmes, just like his plastinates, attempt to capture death in a different way, to freeze time and to preserve the being beyond death. Discussing the wax anatomical models of the eighteenth century, Bronfen argues that 'these models are endemic to a general cultural effort to eliminate the impure state of mutability and decay by replacing it with a pure and immutable wax body double' (1992: 99); arguably von Hagens' models, and by extension his television work, render the decaying body timeless and sanitised in a similar fashion, offering the promise of the 'truth' of the human body and the potential to capture this truth *forever*. The anatomist has said of his work that 'I want to bring life back to anatomy. I am making the dead lifeful again' (Jeffries, 2002: A2). This desire is interesting in relation to the discussion of spectacular television in this book which proposes that we must look beyond the fleeting, distracted gaze to understand television aesthetics. If von Hagens' television programmes are an extension of his plastination project, then television becomes a record of the human body, permanent rather than impermanent, archiving the spectacle of the human body rather than offering passing, fleeting images of it. The spectacle of the human body on television thus challenges the mutability of the body in life. Furthermore, the desire to view the human body as it is presented and explored on television, to see it *in its entirety* at all stages of life and death, is proof of the fact that scopophilia is absolutely endemic to television, rather than being foreign to the medium.

7

The erotics of television

What does it mean to be exploring the erotics of television? As previously acknowledged, scopophilia, or the derivation of pleasure from looking, is a term that can be as equally applicable to television as it has been to film, but here the moments in which that visual pleasure is attached to the structures of desire are explored. Fascination in the previous chapter was often understood as a fascination with extraordinary or abject bodies on television, bodies that the viewer pauses to gawp at, but here fascination is that of attraction and arousal, and the bodies that the viewer pauses to erotically contemplate. The bodies of this chapter are the desirable and desiring bodies located on and at the television screen (and here consideration is also given to what happens when that screen is not a television screen, held at an intimate distance, but a handheld or a movable screen: a tablet or laptop, even a phone screen). In her exploration of 'desired images', those images which viewers 'long to see, and which give back a sense of stimulation or well-being' (2004: 7), Patricia Holland describes, via Barthes (1975), the erotics of the text as 'a static moment [that] interrupts the eager flow of narrative and the incessant demands of understanding. Visual imagery [that] gives us pause. It offers a time out, however short, from rational thought' (2004: 6). Whilst Holland describes the concept of erotics here as relating more broadly to the visual image and its impact on the unconscious, it is her description of a sensual pause, a 'time out from rational thought', which lays open other pleasures in the text, pleasures that might be explored in our understanding of the erotics of television. This chapter will be particularly concerned with the ways in which a variety of contemporary television dramas provide the viewer with intentional erotic spectacle: moments, images, characters, even episodes which both seek to represent and provoke desire. On the other hand, however, it will engage with the flow of the erotic on television, and with what might be termed accidental erotic spectacle. Whilst intentional erotic spectacle might be stumbled upon accidentally, scopophilia can also be located in 'unlikely' places on television; the scopophilic visual pleasures of the medium thus cannot be easily contained in or confined to particular programmes, genres or slots in the schedule. Subsequently, the conclusion of this chapter draws on Barthes' discussion of erotic intermittence (1975) to argue that the appearance and disappearance of erotic spectacle is fundamental to the structures of television narrative and television scheduling. Barthes' exploration of the play between presence and absence in the erotic, and the making of what is 'private', 'public', expresses something of what is specific to a televisual presentation of erotic spectacle.

Research for this chapter began by viewing a series of contemporary dramas in which sex and/or desire play a particularly prominent part, and in which intentional erotic spectacle is foregrounded: for example, *Tell Me You Love Me* (HBO, 2007), *True Blood* (HBO, 2008–14), *Game of Thrones* (HBO, 2011–), *Masters of Sex* (Showtime, 2013–), *Outlander* (Starz, 2014–), *Cucumber* (Channel 4, 2015), *Banana* (E4, 2015) and *Poldark* (BBC1, 2015–). There are of course other texts that could be added to this list, but it offers a snapshot of the increased attention to erotic spectacle in 'high-end' (Nelson, 2007) UK and US long-form serial drama. This programming, and its heavy emphasis on sex and desire, are not accounted for in the critical dismissals of television's visual pleasure explored in the introductory chapter of this book. Casual conversations with viewers enjoying this programming revealed, repeatedly, that viewers' encounters with these series and others frequently focused on the erotic moment. This lead to the conclusion that this engagement required further attention, for as Linda Williams reminds us, in relation to her study of hard core pornography, 'to be moved by pornography is not to be uncritical' (1999: xi). In this context, one may argue that to be moved by the erotic pleasures of television drama is also not to be uncritical about them.

Despite the fact that this work on the erotics of television was inspired largely by the viewing of relatively contemporary television drama, intentional erotic spectacle on television is not an entirely new phenomenon. In the UK, it is easy to think back across a series of such moments throughout the history of television drama: long scenes of Sheila White's barely concealed nudity marking Messalina's sexual conquests in *I, Claudius* (BBC2, 1976),[1] the incestuous lust of *Bouquet of Barbed Wire* (LWT, 1976), the open desire and sexual freedom of *Lady Chatterley's Lover* (BBC1, 1993). Similarly, Amber K. Regis (2012) has charted the history of the 'apparitional lesbian'[2] (and the fleeting moments of lesbian sex on screen) in British TV drama, from *Portrait of a Marriage* (BBC2, 1990) to *Tipping the Velvet* (BBC1, 2002), whilst both Jane Juffer (1998) and Jane Arthurs (2004) discuss the significance of the broadcast of *The Red Shoe Diaries* (Showtime, 1992–7) in the UK and the US in bringing erotic spectacle to television. Thus whilst dramas of desire which dwell on the erotic moment have become more commonplace in the UK and the US since 2010, the concepts and ideas explored in this chapter have a broader historical application than the case studies discussed might suggest. Furthermore, as shall be revealed in the discussion of accidental erotic spectacle in this chapter, encounters with the erotic on television are often deeply rooted in our personal television histories, figured as intensely significant moments in our memories of viewing and often literally returned to through nostalgic viewing or figuratively returned to as part of an 'origin' narrative which explains the formation of our sexual identities.

Textual analyses of the erotics of television are accompanied here by further research into how the ordinary viewer watches and engages with television as a source of visual pleasure tied to the erotic. Viewer research was therefore undertaken, using social media to 'snowball' a questionnaire with a series of open-ended questions about television and desire.[3] Eighty-three extensive and often very candid responses to this survey were received (from 67 women and 16 men[4]); 70 per cent of respondents were based in the UK, and 16 per cent in the US. Just over 60 per cent of those participating in the research were in their thirties and forties; just under a quarter were aged between 50 and 75. The results of this survey, which showed some striking trends and discourses running through the respondents' articulation of their erotic relationship with television, will be discussed below. This research both concurs with, extends and sometimes contradicts Hazel Collie's (2014; 2016) longer-form 'life story' style interviews with women about their historical engagement with television, which also revealed television as a significant site of desire. Significantly, Collie's work has called for a greater engagement with the erotic on television: she argues that 'we do not recognize the different space that television creates for desire' (2014: 160) and concludes that 'the domestic should be seen as [a] space for attraction and desire to be played out' (ibid.: 172). This chapter provides an exploration of this. Further, and leading on from Collie's work, particularly striking trends in women's responses to this survey will be discussed below and coupled with an analysis of programming which very clearly addresses and represents female desire. This chapter thus picks up the critical strands of film and television scholarship of the 1980s and 1990s which sought to investigate the female gaze and will offer some analysis of how this gaze is particularly articulated in contemporary television drama.

The proliferation of television sex

In Linda Williams' exploration of hard core pornography, she discusses, via Foucault, the 'modern compulsion to speak incessantly about sex' (1999: 2), and the way that this has manifested in the proliferation of the audio-visual representation of sex in the pornographic film. Whilst the texts Williams discusses do indeed represent an increase in the filming of explicit sex, they must be sought out, tracked down and purchased (or downloaded), and do not, therefore, appear everywhere in the media flows of everybody's everyday life. On the other hand, however, television's increasingly widespread presentation of sex might arguably be seen to even more effectively represent the compulsion Williams (and Foucault) speak of. This proliferation of sex on the most ubiquitous of media is carefully documented in Jane Arthurs' work on television and sexuality (2004) as well as being explored in Williams' later writing (2008); indeed, Arthurs finds sex in television across multiple channels and genres (television drama, documentary, factual entertainment programmes and television news) and her work outlines the nature of sexual citizenship and sexual consumerism as it relates to television.

The writing that documents the proliferation of sex on television also sits alongside critical literature that has commented on and critiqued the 'pornogrifying' of culture more broadly, that is, the infiltration of a 'porn aesthetic' into the mainstream and into everyday life. Brian McNair's concept of 'porno-chic' (2002) describes the commodification of desire and the promise of sexual (visual) pleasure in the late twentieth and early twenty-first centuries, proposing that 'sex is about money, and never more so than in the capitalist societies of today' (ibid.: 6). McNair goes on to argue that this commodification is particularly evident on television, discussing Channel 4's 'Red Light Zone' in the UK, a late-night schedule slot in which 'risqué' programming was shown, and the ways that sex was presented as a marketing strategy on this channel in the mid-1990s.[5] He suggests that these programmes are not pornography in themselves but 'porno-chic', 'referring to pornographic texts in their content and style, but not singularly dedicated to the task of sexual arousal in the manner of true porn' (ibid.: 86), and proposes that they form a part of the 'economically driven sexualisation of television' (ibid.: 95). Arial Levy describes a similar proliferation of sexual imagery across the public sphere in her work on US 'raunch culture', which she also implicitly relates to her own television viewing:

> I first noticed [raunch culture] several years ago. I would turn on the television and find strippers in G-strings explaining how best to lap dance a man to orgasm. I would flip the channel and see babes in tight, tiny uniforms bouncing up and down on trampolines.
>
> (2005: 1)

Levy points towards what she calls 'harem themed reality shows' (*The Bachelor* (ABC, 2002–), *Who Wants to Marry a Millionaire* (Fox, 2000), *Joe Millionaire* (Fox, 2003), etc.) and their concomitant pressure for women to be overtly and publicly sexual as examples of the role that television plays in the normalisation of raunch culture. Whilst the dramas under discussion in this chapter are not taken as further examples of porno-chic or raunch culture – they neither mimic a porn aesthetic nor share its 'dedicated task' – these histories contextualise the turn towards explicit sex on television and provide further evidence of the proliferation of sex on television beyond the 'high-end' drama. However, whilst, as Arthurs argues, 'genres addressed to high-status audiences are allowed to be more explicit and controversial' (2004: 24), her work, and the research of McNair and Levy, also suggest that the proliferation of sex on television extends beyond the generic boundaries of the quality drama.

It is beyond the critical reach of this chapter to offer a detailed analysis of television pornography, though arguably a study of visual pleasure and the erotic on television must acknowledge that pornography offers one of television's more obviously *visual* pleasures. Arthurs and James Aston (2012) both acknowledge that television is an important site for pornography even though this genre of programming is infrequently acknowledged in the scholarship on the medium. Aston's work offers the most sustained analysis of television pornography; analysing pay-per-view TV porn and the content of subscription channels, he offers a detailed history of this genre. Aston's thesis is that in comparison to more explicit and misogynistic versions of the same narratives in online pornography, TV porn is more concerned with depicting female desire and less concerned with showing the most explicit images of penetration (he describes these programmes as a 'frenzy of the invisible' (ibid.: 84)). His analysis highlights how, in his chosen texts at least, 'women [are situated] as more active characters who seek out sexual intercourse. The result is a more consensual and democratic sexual relationship' (ibid.: 85). Whilst one might wish to argue that the programming Aston discusses partially conceals the sexual exploitation of women or their degradation which is more fully revealed in the online edit of the same film shoot on the multi-media porn producer Television X's website, the plethora of soft core porn channels available on multichannel digital platforms are interesting to this study because they are further evidence of the domestication of sexual imagery via television. As Arthurs suggests, there is no longer 'such a clear divide between the respectable [television] audience […] and the despised "dirty mac" brigade of sad misfit men that formed the stereotyped consumers of pornography until relatively recently' (2004: 46). This idea is also borne out in the audience research at the heart of this analysis of the erotics of television. Of the respondents to the Television Desire Survey at the centre of this research, 57 per cent reported having consumed porn at home on their TV (or TV-replacing) screens, whilst just under 50 per cent of the female respondents and just under 20 per cent of the men had never watched any pornography via their television. Whilst most people surveyed had consumed pornography at home via video or DVD, or through a non-subscription TV channel (only two people responding to the survey had used the subscription services of pay-per-view films Aston discusses), pornography was also viewed at home via the convergence of television and the internet (streamed through a laptop and onto a TV screen, or viewed on a laptop, iPad or phone whilst at home), as well as on sex chat phone-in channels and, sometimes accidentally, as part of television's flow. This research suggests that, as Arthurs argues, pornography has been simultaneously domesticated and normalised via television, and that television has become a conduit for erotic contemplation, again beyond the bounds of the quality drama on mainstream channels and more 'reputable' subscription networks. Pornography in the home has therefore become part of the fabric of television watching, and thus fairly regularly sits alongside the medium's other visual pleasures in an ordinary, privatised, domestic space.

Whilst the work discussed above situates the erotic and the spectacle of sex at the centre of popular entertainment and at the pornographic margins of the electronic programme guide, there has also been a marked proliferation of sex and desire in the popular serial television drama. Certainly, an emphasis on erotic spectacle has become a feature of the 'quality' US dramas produced by American premium cable TV channels like HBO, Showtime and Starz: Marc Leverette argues that 'by turning its everyday content into the spectacular and aestheticizing the taboo, HBO is an example par excellence of what Caldwell [...] termed "boutique television"' (2008: 141). The most prominent recent example of this trend is the epic fantasy series *Game of Thrones*, in which sex is frequently represented as both an everyday activity, a spectacular visual pleasure within the diegesis and as a significant aspect of the power plays at the heart of the saga. Characters are featured having sex out of duty and for control or influence, as well as out of lust and desire, and the series has been widely criticised for its heavy use of nudity and its depiction of sexual violence. The programme's depiction of rape has been repeatedly condemned,[6] and beyond this, there have been prominent critiques of the gratuitousness of its sexual imagery (see Jones, 2014; Siede, 2015). Indeed, the blogger Myles McNutt (2011) coined the now widely used term 'sexposition' to refer to the ways in which vital plot exposition in *Game of Thrones* is often given against a 'backdrop' of sex or nudity (as seen in the season one episode 'You Win or You Die'), leading to 'charges of exploitation on the part of the writers, or an underestimation of the viewer's capacity to concentrate on complex expositionary dialogue without a visual "pay-off"' (Wheatley, 2015a: 61). To focus in on an exemplary episode, this visual 'pay-off', a spectacular visual pleasure detached from the evolution of narrative, is seen earlier in the season one episode 'The Wolf and the Lion' in a scene at the end of the episode where Ned Stark (Sean Bean), the Hand of the King, confronts Master of the Coin Petyr Baelish (Aiden Gillen) about the King's progeny in a brothel. Whilst the dialogue of the two men can be heard off screen, Stark's guard is shown staring at a topless prostitute positioned in mid shot as part of a shot-reverse shot sequence at both the beginning and end of the scene. Her image, and particularly her lascivious gaze and her nudity, transfix him, and, arguably, the viewers: the scene repeatedly cuts back and forth between his staring face and her exposed torso. This sequence thus *initially* seems to confirm the fact that women are depicted as the object of a particularly rampant form of scopophilia in the series, and that they are frequently shown to be (gratuitously) naked. However, *Game of Thrones* has been careful to offer some gender equality in its depiction of naked and sexualised bodies. An earlier scene in this episode – which begins with a close-up of the castle prostitute, Ros (Esma Bianco), being fucked standing up and from behind by the young ward of the Stark family, Theon Greyjoy (Alfie Allen) – exposes both characters equally: as they finish, both actors are shown naked in a full-frontal medium long shot (see Fig. 7.1).[7] Here, the dialogue that follows, in which Theon's relative lack of power and status is revealed by Ros' taunting, again shows sexually explicit action as a backdrop for, or precursor to, narrative exposition. However, the scene explicitly aims for gender equity in its presentation of the naked body. Furthermore, there is a representational tension in this sequence, which both spectacularises the sexual organs of the two actors whilst challenging its viewers to read Allen's penis and Bianco's breasts as ordinary or unspectacular. Part of the point being made here, then, is that these characters are so comfortable

Fig. 7.1 The ordinary spectacle of sex in *Game of Thrones* (HBO, 2011–)

in their nudity that they can continue a conversation that is not *about* their sexual relationship or their bodies whilst naked; thus the viewers are challenged to look both at but also beyond the nudity on screen here.

To dwell on this particular episode further, a tender moment in which the brother of the King, Renly Baratheon (Gethin Anthony), has his chest shaved by his lover, Loras Tyrell (Finn Jones), focuses visually on two attractive young men being intimate with each other whilst, on the other hand, the dialogue in the scene speaks of Renly's desire for more power and greater respect from his brother. The scene also reveals the King's lack of desire for his rich wife. As he complains that Loras is a better swordsman in combat than him, a cut is made to a seemingly unmotivated close-up of his nipple, and, initially, this shot seems to be entirely presented as erotic spectacle; it 'interrupts the eager flow of narrative and [...] gives us pause' (Holland, 2004: 6). However, as the razor scrapes around this most sensitive of areas, we might also understand that a narratively significant point is being made here about coping with and surpassing a sense of vulnerability; that in order to gain power, one must learn to deal with weakness. Here the shot of an intimate body part has thematic resonance and thus cannot subsequently be identified as pure erotic spectacle: it serves narrative purpose as well as providing visual pleasure. Indeed, even in the sequence in the brothel discussed above, the gaze of the knight on the topless woman does actually underscore the fact that the King's lasciviousness (and the resulting large number of his bastard children) is being discussed, and that this will have a significant impact on the unravelling of the Baratheon family's grip on their rule. The depiction of lust thus carries thematic significance even whilst it might be seen as a distraction from dialogue and narrative exposition. These examples thus reveal the difficulty of identifying 'gratuitously' erotic imagery in this series. Whilst there is clearly a way in which to enjoy these moments of the erotic for their own sake, one could also argue that sometimes even images of naked torsos and erect nipples can have meaning beyond their status as images to be looked/gazed at.

Masters of Sex, the Showtime series about the sexual research of William Masters (Michael Sheen) and Virginia Johnson (Lizzy Caplan) also draws our attention to the propriety of sexual openness and the viewing of explicit sex, though these issues are more clearly, and self-consciously, explored through the narrative of the programme. The programme focuses on Masters and Johnson's struggles to legitimate their study, to find an appropriate and ethical methodology for the research and to negotiate their own roles in their work and their attraction to each other. Much of the drama (particularly in the first season) revolves around them watching people having sex or sexually stimulating themselves (in their lab, or in a nearby brothel when their research is forced 'underground'), as well as Masters and Johnson's real and imagined sexual activity with each other and other people. As the series progresses, the revelation of female desire and arousal as a real and measurable experience becomes key, and much of the drama in *Masters of Sex* is found in the need to have that desire satisfied. Virginia Johnson is positioned at the centre of the narrative as a sexually liberated figure that is able to take pleasure in sex and the viewing of sex. She is an autodidact who challenges the status quo in the hospital and in the field of sexual research despite starting out at the hospital in a secretarial role; when discussing the research with Jane (Heléne Yorke), another secretary at the hospital and one of the research subjects, Virginia says of Freud's attempt to theorise female sexuality, 'I'm not so sure I care what a male psychologist has to say about female sexuality'.[8] This is an example of the ways in which Virginia is figured as 'taking charge' of her own sexuality, and also as seeing the absolute (political) necessity of revealing the 'truths' of female sexuality more broadly.

Bill Masters, on the other hand, is an interesting figure in relation to the repression of the desire for sex, and for watching sex. In scene after scene, the denial of his spectatorial pleasure is the key tension in the scene; he is a repressed lover who can barely bring himself to make love to his wife, and remains a largely impassive viewer watching women masturbate or couples making love, constantly denying his wife's (Caitlin Fitzgerald) suggestion that that 'When you do watch, it must be sexy?' Through Masters, the series explores the tension between involvement and detachment in the position of the viewer: his position as a doctor in the field of sexual research necessitates his detachment of course, though this position draws a tension between public and private articulations of desire. Sheen's portrayal of Masters hints at this repression of desire (his frequently clenched jaw, the tension in the eyes framed by a studiedly impassive face) and the use of psychic point of view or daydream sequences occasionally allow Masters' feelings of desire (for Virginia Johnson) to be (briefly) interjected into the narrative. Without Virginia (she is fired or quits at various points in season one), Masters struggles to make connections with his subjects and thus she is frequently configured as repurposing a gaze on the spectacle of sex that is problematic when it is associated with the doctor alone. (He both doesn't *feel* enough *for* his subjects, he disassociates and also, as arbiter of the male, medicalised gaze is potentially too controlling of his subjects in his viewing.) His gaze is figured as openly desiring only when he is looking at Virginia. The series is thus attentive to the issue of placing women's bodies under a scrutinising medical or sexual gaze and Virginia repeatedly recontextualises this gaze in the context of sisterhood or shared experience. In episode one, 'Masters of Sex', for example, Virginia persuades Jane to participate in the research by delivering this speech: 'It's a whole new world we're opening up. Groundbreaking. Very exciting – for women especially. It will probably be the biggest change to women's lives since the right to vote.' Following this scene, and a brief sequence in which Jane is shown signing up for the study, her participation in the research is cross-cut with Virginia's description of the feeling of the female orgasm. Here close-ups of Dr Masters applying probes to Jane's naked torso, and then close-ups of her face and breasts as she masturbates, are cross-cut with Virginia describing her orgasm to the doctor, with this narration continuing across both scenes. Rather than a solely dissociative, objectifying (male) gaze then (the scene as seen from Masters' point of view, perhaps), the edits try to ensure that the spectacle of Jane's desire at this moment might be read as an intimate depiction of shared (female) sexual experience.

The examples discussed above represent television sex as a strategy of distinction, as a marker of televisual exhibitionism (using Caldwell's (1995) mobilisation of this term), as extraordinary or edgy *and* spectacular in the sense that these programmes dwell on moments of desire and the representation of sex as offering a form of 'heightened' visual pleasure. They both, however, also radically seek to present sexual activity as ordinary, everyday, as part of the human experience (hence the continuation of conversation during the sex scenes discussed in *Game of Thrones* which tells us that life goes on around, despite or even during spectacular sex, or *Masters of Sex*'s insistence that sex is natural and essential to human happiness).

This ordinariness can be related to the idea of sexual realism in television drama. A significant proportion of women who responded to the Television Desire Survey argued that realism was *central* to their enjoyment of television sex, either in terms of depicting the sometimes awkward or difficult/unromantic realities of sex (a comment made by just over 20 per cent of female respondents) or the realistic occurrence of sex within a developing narrative (as discussed by just under 20 per cent of female respondents). For example, a 34-year-old, female, bisexual events co-ordinator from the US stated:

> Most of the time, I don't like televised sex. There are a few reasons for this. Primarily, I would say that I find of lot of it gratuitous and unrealistic. Network sex – the keep-your-bra-on sex – is little more than making out, and Pay Cable sex is often just boobs for the sake of boobs, and there isn't a whole lot in between.

Whilst this viewer identifies most US TV drama as lacking in sexual realism, a number of key dramas have striven for a sexual frankness which takes a 'warts and all' approach to sexual realism; here the spectacle of sex is tied to the portrayal of recognisable experience. In the US, the serial drama *Girls* (HBO, 2012–) takes this approach; sex in *Girls* is frequently messy and embarrassing and is depicted with an intimacy that encourages us to identify and engage with these feelings.[9] In the UK, the writer-producer Russell T. Davies has also become associated with sexual realism, both in relation to his explicit depiction of gay sex in *Queer as Folk* (Channel 4, 1999–2000) and more recently in his broader depiction of queer and straight sexualities in *Cucumber* and *Banana*. Speaking in the online documentary series that accompanied the latter two series, *Tofu* (4oD, 2015), the actor Julie Hesmondhalgh argues that 'the sex that you see [in Davies' dramas], it's sometimes beautiful and really sexy, it's sometimes disastrous, it's sometimes messy and it sometimes doesn't happen at all'. For Davies it is clear that this realism is tied to a public service responsibility to depict on television the realities of sex, particularly for young people, rather than, or as well as, producing realistic sex that viewers might find visually pleasurable. This idea of sexual realism as a public service, or what Barrie Gunther describes as television's role in sexual pedagogy (2002), is explored in the documentary series *Tofu*, in the episode 'Queer as F**k'. In this episode, the actor Andrew Hayden Smith recounts the experience of watching *Queer As Folk*: 'I loved [it]. I mean it was a huge part in dealing with my sexuality', and Gary, a charity volunteer, recounts that

> at the time I hadn't seen any gay sex and I remember getting off on it [...] Seeing other gay people on TV did make a difference to me and it was almost like I could feel accepted and wasn't an outcast any more.

The significance of television in relation to the idea of 'sexual awakening' or the formation of sexual identity was also strongly present in the responses to the Television Desire Survey. Without being specifically asked about when they first felt 'attraction' in relation to people or characters on television, a quarter of the survey respondents argued that television had become associated with desire for them in childhood, and this figure was much higher for bisexual women (36 per cent), lesbians (40 per cent) and gay men (75 per cent).[10] As a straight female librarian in her twenties put it, 'I think watching sex on television has been an important part of helping me to explore and try to understand sex as a young girl/teenager'. As with Collie's (2014) study, often this attraction was expressed in relation to the figure of the 'crush', someone who is admired from afar but without the potential for a real sexual relationship. For example, a straight man in his thirties from the UK remembered:

> My first crush was Scorpio from the 90s series *Gladiators*. I remembered being quite captivated with her whenever she was onscreen but, as it was a family show that I used to watch with my parents and siblings, feeling quite awkward about it. I used to videotape the episodes and watch them back privately, keeping the shows where she'd feature prominently. I ended up keeping those tapes for many years until I finally dumped my VHS player as it became obsolete. I'll still look for the shows on YouTube every now and then to remind myself.

Here the memory of desire for (or a 'crush on') a televisual para-sports star is intertwined with memories of 'live' family viewing and private time-shifted, repetitive, serial viewing. The memory of desire is thus intertwined with feelings of discomfort around family viewing and the pleasures of a closer, more singular relationship with television where the TV set and the VCR allow for a private erotic encounter to take place. Indeed, it is not a great leap from this account of adolescent television viewing to Linda Williams' exploration of the 'electronic interactivity' of the porn video:

> Although home viewing of a video, remote control in hand, gives a viewer the freedom to isolate and replay certain scenes, the pornographic 'instant replay' has the added dimension of eroticizing the act of manual control in a potential context of privacy that is ideally suited to masturbation.
>
> (1999: 304)

The above account thus speaks of the intimacies of televisual desire that have been largely ignored in the critical work on the erotic in screen media.

The proxemics of television – televisual desire up close

No account of television's visual pleasure as it relates to the scopophilic or desiring gaze could proceed without engagement with Laura Mulvey's essay 'Visual Pleasure and Narrative Cinema' (1975), which laid the foundations for the analysis of gendered looking relations in film and, subsequently, television studies.[11] Whilst earlier work on television form argued for the non-applicability of Mulvey's arguments about voyeurism, the fetishising gaze and scopophilia to television, it is proposed here that her work might be used as a critical springboard for the theorisation of televisual desire, and that Mulvey's account of the intensity of the spectatorial engagement with the filmic image needs to be adapted, not abandoned, when it comes to thinking about our erotic engagement with televisual spectacle. The rejection of Mulvey's characterisation of the voyeuristic (male) gaze as the dominant spectatorial position in relation to television is first articulated in John Ellis' early work on the medium. He argues:

> Broadcast TV's lack of an intense voyeuristic appeal produces a lack of the strong investigatory drive that is needed alike for tightly organized narration and for intense concern with the 'problem' of the female. Similarly, the regime of broadcast TV does not demonstrate a particular drift towards fetishistic activity of viewing.
>
> (1982: 142)

Ellis goes on to qualify this by saying that:

> the fetishistic regime does operate to some extent, however, as does voyeurism. Its characteristic attention in broadcast TV is not directed towards the whole body, but to the face. The display of the female body on TV [...] is gestural rather than fascinating.
>
> (ibid.)

However, he largely dismisses the idea that the visual pleasure of television has the intensity of a comparative encounter with cinema. Similarly, John Fiske engages more directly with Mulvey's work in his rejection of its applicability to television, concluding that 'television is more interactive than voyeuristic' (1987: 226) even whilst his qualification is that television advertising fragments and fetishises the female body. Sandy Flitterman-Lewis (1992), in her account of the applicability of psychoanalytic theory to television, fails to move beyond the primacy of 'glance theory' (as first articulated in Ellis 1982) in characterising a purely casual relationship between viewer and screen, as does Suzanne Moore (1998), and thus these authors do not see the applicability of Mulvey's work to an understanding of television's looking relations. However, all of this work fails to account for the presence of intended or accidental erotic spectacle on television, spectacle that provokes a deeply scopophilic, non-casual form of viewing, which might indeed be voyeuristic on some level. For example, the young man's memories of television viewing discussed above, which describe an intense and voyeuristic relationship with a 'crush' on screen and which recall the utilisation of a domestic viewing technology (the VCR) and an online archive (YouTube) in order to intensify or revisit this televisual spectacle, cannot be accounted for in this dismissal of Mulvey's work.[12] This viewing, as representative of a whole range of discussions about television viewing and desire in the survey at the centre of this chapter, precisely fits Mulvey's description of scopophilia as 'taking other people as objects, subjecting them to a controlling and curious gaze' (1975: 8). Whilst the 'controlling and curious gaze' has previously been explored in relation to the medicalised body on television in the previous chapter of this book, here an intense curiosity is found in relation to the spectacle of desire on television. Mulvey argues that 'structures of fascination [in cinema are] strong enough to allow temporary loss of ego [...] (I forgot who I am and where I was)' (ibid.: 10); in contrast, one might argue that television's structures of fascination do not result in this temporary loss of ego but rather a fantasy of total surveillance (not 'I forgot who I am or where I was' but 'I want to see every bit of you', perhaps). Television's potential to be immersive leads to an intensity which is found in the medium's proximity, rather than the awed and enraptured distance of the cinema spectator to the screen, and thus to a different form of fascination.

Television's intimacy with images and figures which are simultaneously domesticated (brought into the intimate sphere) and absent (broadcast from another place, and often from another time) is also (partially) described in some of the work which takes up Mulvey's critical challenge in relation to cinema. For example, in Mary Ann Doane's description of cinematic voyeurism she stresses that 'distance between the film and the spectator must be maintained, even measured' (1982: 77), and argues that 'all desire [...] depends on the infinite pursuit of its absent object' (ibid.: 78). In reference to Noël Burch's (1973) work, Doane argues that 'the viewer must not sit either too close or too far away from the screen. The result would be the same – he would lose the image of his desire' (ibid.). Perhaps it seems odd to be using Doane's argument to propose that television might also be seen as a medium of the voyeur, given our fundamental closeness to its screen/s? However, if the work discussed above (Ellis, Fiske, et al.) understands television as a screen too close to its viewer to produce the necessary distance for voyeuristic contemplation, it misses a trick. In fact, the television image is simultaneously intensely close but also always held at a distance, and, through the broadcast or streamed image, the object of desire is simultaneously physically close to us but also always tantalisingly absent, 'beamed in' from another place, and often another time. In fact, for a number of the respondents to the Television Desire Survey, an increasing closeness of television to the viewer and the viewer's body *intensified* their voyeuristic erotic engagement with the text, enabling them to gaze intently at the simultaneously present and absent bodies on screen. Respondents were asked to consider the impact of watching sex on television on different screens (TV set, laptop, tablet, mobile phone, etc.), and whilst just over half of the respondents stated that what they watched on had no impact on their erotic engagement with television,[13] of those who expressed a preference for a particular screen, 38 per cent preferred to view on a handheld tablet or phone screen and 21 per cent preferred to watch on a laptop. Whilst for some people a handheld screen provided greater privacy ('seeing as I have small children, most of the sexy stuff I watch is on my iPad';[14] 'The enjoyment is indirectly linked to the device not because of the laptop itself but because the laptop indicates more privacy which can lay the ground for more "engagement"'[15]), other people commented on the ability to take the handheld screen into intimate spaces in the house ('I think that watching television on my kindle or iPad in bed makes things feel more erotic'[16]) and also on the increased intensity of viewing and listening to an entirely privatised text ('I think the iPad has a stronger impact on erotic engagement simply because I use headphones. It's easier to feel more engrossed in the TV when outside distractions/noises are removed'[17]).

It is interesting that those preferring to watch programming with sexually explicit, erotic content on handheld devices were exclusively female; whilst the particular engagement with a female (desiring) gaze on television will be explored at greater length below, there is something to say here about the ways in which the very close and privatised view of erotic imagery afforded by watching on a tablet particularly appeals to women.

This is certainly in line with the historical trajectory of women's pornography and erotica that is explored in Jane Juffer's work, which traces a history of women's literary erotica, internet-based pornography and the impact of cable TV and the VCR on women's consumption of erotic and pornographic texts. In this work, Juffer argues that women have 'carve[d] out spaces for [pornography and erotica's] consumption and for fantasy within [their] daily routines' (1998: 5) and within the domestic sphere. Following Juffer, Jane Arthurs proposes that 'the development of new technologies – the video recorder, cable television and now the internet – [has] changed the material conditions of domestic space and its temporal rhythms, carving out new possibilities for women as consumers of pornography' (2004: 51). The viewer research discussed above suggests that the privacy and intimacy afforded by the tablet screen have further enabled women's consumption of increasingly explicit television programming, an argument that was made earlier in the decade in relation to women's erotic literature (specifically the E. L. James *Fifty Shades of Grey* (2011) trilogy) and the e-reader (see N. Richards, 2015). The same private, intimate space is created between viewer and tablet and reader and e-reader: a space that can not only shield her from public scrutiny of her choice of television programme or novel, but also creates an intensely intimate space for engagement with erotic fantasy. The viewer who discusses the impact of wearing headphones on her erotic engagement with television (quoted above) thus succinctly characterises this intensely private spatial relationship. Perhaps this desire to pull closer to the screen can also be explained by Walter Benjamin's description of the 'desire of the present-day masses to "get closer" to things' and his observation that 'Everyday the urge grows stronger to get hold of an object at close range in an image, or, better, in a facsimile, a reproduction' (1999: 105). For television, the tablet offers this 'getting closer' to scenes of intimate, erotic spectacle.

Mapping female desire and the erotic spectacle of television

Rather than being a textually unified or homogeneous medium, television is characterised by its textual diversity and the heterogeneity of its programming. This not only refers to the diversity of television genres which make it difficult to generalise about what television *is*, but also to the different viewing positions that are created within this diverse programming. Television thus offers the potential to provide 'something for everyone' when it comes to the presentation of desire and the articulation of gendered/sexed looking relations. This heterogeneity certainly raises a challenge for those wishing to explore Mulvey's (1975) essay in relation to the medium. One of her core arguments was that 'unchallenged, mainstream film coded the erotic into the language of the dominant patriarchal order' (1975: 8) and that, concurrently, 'woman displayed as sexual object is the leitmotif of erotic spectacle [in the cinema]' (ibid.: 11). Whilst Mulvey's (over)statement of the monolithic nature of the male gaze and female 'to-be-looked-at-ness' in cinema has been challenged in film scholarship, in relation to television, as a far more heterogeneous medium, it is even more problematic to make such claims on firm ground. Jane Arthurs (2004), amongst others, has shown very clearly that queer sexualities and subject positions have become increasingly visible on British television (particularly in UK TV drama), and Jane Juffer has argued that 'the focus on women as sexual agents, especially on the [US] premium channels, usually overwhelms any attention given to men', and that 'in this sense we can see that adult cable programming shares many of the conventions of women's print erotica' (1998: 201). Juffer does go on to qualify this by stating that 'sexual mobility' is generally reserved on television (in the late 1990s at least) for 'white, straight, upwardly mobile women' (ibid.: 202). However, what is clear in this work is that whilst one might wish to transpose some of Mulvey's ideas about an intense, voyeuristic or scopophilic relationship between viewer and text to television, the notion of the (implicitly heterosexual) male gaze is less useful in this context. This section of the chapter is concentrated most fully on erotic spectacles on television that largely call out to, or focus on, female heterosexual desire. This is not just because straight women constituted the largest group in the Television Desire Study, but also because of a recent increase in interest in straight female viewers and their consumption of television's erotic spectacles, partly inspired by the popular success of a series of television dramas (particularly the BBC costume drama *Poldark* and the fantasy-historical drama hybrid, *Outlander*, on the US premium network, Starz). The issue of the televisual female gaze thus seems difficult to ignore in the contemporary moment.

In her engagement with the figure of the female spectator, Mary Ann Doane turns towards television to describe women's inability to disconnect from the text and to therefore produce a fetishising gaze on the figure of the man: 'The association of tears and "wet wasted afternoons" (in Molly Haskell's words) with genres specified as feminine (the soap opera, the "woman's picture") points very precisely to this over-identification, this abolition of distance, in short, this inability to fetishise' (1982: 80). This figures the female spectator of television as 'over-involved', unable or unwilling to take fetishistic visual pleasure outside of narrative engagement, and this was sometimes borne out in the viewer research at the centre of this chapter. Comments such as 'I found some of the *Game of Thrones* characters attractive [...] because it captured my imagination so strongly and I got so engrossed in it; this seemed to accelerate my feelings towards [these] characters'[18] suggest an engaged spectator who is more enraptured by the unfurling of narrative than she is by the sight of the figures on screen. Respondents discussed the importance of ongoing seriality in the development of their desire for a character ('It's as I get to know each character that the attraction develops'[19]), or for their continuing enjoyment of a series ('I think having a "love interest" that you follow is part of what gets you hooked'[20]), which confirms Collie's comparison of television desire with the 'long term relationship' (in contrast to cinema's 'nerve wracking and potentially one-off first date' (2014: 173)).

Comments about seriality and the slow 'unfurling' of desire in relation to character development in a serial narrative were particularly made in response to the question 'Are you more likely to be attracted to a fictional character, a television presenter, a public figure (e.g. politician, sports personality) or a 'real person' (e.g. reality show/documentary participant) on TV?' There was a starkly marked gender difference in the answers to this question: whilst just under 80 per cent of all women said they were primarily attracted to fictional characters on television, just over 80 per cent of all men said they were equally or more attracted to 'real people' on television, and all of the gay male respondents said they did not differentiate what type of person, fictional or real, they related to on television in terms of attraction. Commentary on this within the survey concentrated on the fact that for the female respondents, television had become, or always was, a medium of fantasy, an escape from disappointing realities of 'real' romantic attachments ('I'm much more likely to feel attracted to a fictional character – real people have a habit of ruining my perception of them'[21]), or a springboard for further erotic fantasy ('I don't write fanfic, but I like to imagine stories akin to fanfic for these characters. If it's a "real" person like an actor or TV presenter I don't like reality to intrude on my imagination'[22]). Female respondents thus talked about escapism of a television 'crush' as well as the 'safety' of developing an attraction to a fictional character on screen one could never actually meet.

The figure of Doane's female spectator caught in the grips of over-identification with the characters in a romantic narrative is pointedly represented in the *Masters of Sex* episode 'All Together Now'. Following the revelation that she is living a frustrated life in a sexless marriage and has never experienced an orgasm (we learn during the first series that her husband is a closeted homosexual), the provost's wife Margaret (Allison Janney) goes to the cinema to see *Peyton Place* (US, 1957). In this scene we see a shot-reverse shot sequence where cuts are made between a slow zoom of Margaret isolated in the cinema and the image of two lovers bickering on screen. The intensity of this cross-cutting sets out to express exactly the relationship between viewer and screen described by Doane. During the slow zoom in on her sad face, we hear melancholic extra diegetic music (piano and strings) as well as the off-screen dialogue 'I kissed you, you kissed me, that's affection, not carnality, that's affection, not lust, you ought to know the difference'; this soundtrack thus speaks of Margaret's frustrations and thwarted desire and her identification with the characters on screen. Cutting back to her point of view shot of the screen thus visually articulates her romantic engagement with, or rather entanglement in, the text (which is underscored by the fact that we have earlier seen her reading the novel of *Peyton Place* in bed); as the arguing characters kiss, a cut is made back to Margaret blinking and biting her lip, representing kinaesthetically the bodily reactions of this thwarted desire. *Peyton Place* is a significant choice of text in a programme so engaged in a televisual exploration of women's desire: as it went on to become a television soap opera (ABC, 1964–5), it represents exactly both kinds of text Doane discusses in her article on female spectatorship. However, what follows this scene for Margaret is a sexual awakening: she bumps into the young Dr Langham (Teddy Sears) outside the cinema, and their subsequent brief but passionate affair refigures her as a desirable *and* desiring woman who takes (visual) pleasure in looking at her young lover. She thus quickly moves on from the intensity of the over-identification with the characters on screen in her viewing of *Peyton Place*, and takes ownership of a different kind of gaze.

In the Television Desire Survey, both straight and gay women more frequently discussed the dual pleasures of narrative *and* erotic spectacle:

> In fact, a lot of times when I am watching television there is a dual enjoyment going on: first off it's the emotional structure and the plot of the show, but secondly, if there is someone I find attractive, then watching them on screen is a major part of that enjoyment.[23]

Furthermore, there was strong evidence throughout the responses that both male *and* female viewers felt strong attraction to, desire for and arousal based on the figures that they saw on screen. Typically, respondents of both genders discussed their multiple attractions to figures they encountered on television ('I'm constantly attracted to people on television. I can't really remember a time when I wasn't'[24]; 'I have been attracted to many television characters. I think [there] may be too many to count [...] Desire, or at least attraction to characters, probably plays an important role in what I watch'[25]). This discovery of gender equity in the erotic visual pleasures of television thus speaks to E. Ann Kaplan's call in the early 1980s for a greater understanding of female desire and her call for films that 'satisfy our craving for *pleasure*' ([1983] 2010: 212) and Steve Neale's engagement with D. N. Rodowick's work in which he argues that 'the male image can involve an eroticism, since there is always a constant oscillation between that image as a source of identification, and as an other, a source of contemplation' ([1983] 1992: 281). Whilst Kaplan, Neale and Rodowick were specifically contemplating women's engagement with erotic spectacle in the cinema, it is clear that in contemporary television drama this contemplation of erotic spectacle also takes place. If, as Linda Williams has argued, 'for women, one constant of the history of sexuality has been a failure to imagine their pleasures outside of a dominant male economy' (1999: 4), then perhaps an examination of television will offer precisely this engagement with female (visual) pleasures.

Collie has also argued that television is a privileged site for the articulation of female desire. Through her interviews with women at various life stages, she pointedly concludes that 'It is clear from their comments [about the objects of their televisual desire] that television is not always glanced at' (2014: 169). Lorraine Gamman and Margaret Marshment ask a series of significant questions about this in their 1998 collection *The Female Gaze*, though in this work, Gamman (1998) and Budge (1998) concentrate largely on the female viewer's gaze on the woman on television as the source of identification (Gamman: 16) and/or desire (Budge: 104). Geraghty (1981) and Fiske (1987), on the other hand, have also both explored the soap opera as a genre that 'show[s] and celebrate[s] the sexuality of the middle-aged woman, and thus articulate[s] what is repressed elsewhere on television as in the culture generally' (ibid.: 184). Fiske's argument particularly resonates with the exploration of visual pleasure in this chapter when he claims 'All soaps are highly sexual, and many women use terms more conventionally applied to male pornography to describe their reaction to them' (ibid.: 185). Whilst Fiske acknowledges that press coverage of the soap opera frequently emphasises discussion of love scenes or the depiction of the bodies of male 'hunks', he ultimately concludes that 'it is relationships, rather than the body of the male, that are the source of the erotic pleasure offered by soaps' (ibid.: 186). He therefore proposes that secondary texts such as soap magazines 'often emphasize the physical, compensating for the soaps' refusal to sexualize the body of the male in a feminized inflection of voyeurism' (ibid.: 187). This oscillation between textual and extra-textual presentations of erotic spectacle is important; whilst in Fiske's soap opera example the presentation of the male as the bearer of 'to-be-looked-at-ness' is figured differently in the soap and its secondary texts, in the costume drama, as explored below, the secondary text (review, preview and, more latterly, social media posting) has worked to activate, explore and celebrate a fetishisation of the male body more present in the text itself.

In the UK, the BBC's 1995 adaptation of *Pride and Prejudice* has often been seen as something of a watershed moment for the representation of female desire on television. The much discussed, parodied and revisited moment in which Colin Firth as Mr Darcy steps dripping wet out of pond to be ogled by Elizabeth Bennett (Jennifer Ehle) (and, it is presumed, the viewers at home) is referred to in Lisa Hopkins' work when she argues '*Pride and Prejudice* [...] is unashamed about appealing to women – and in particular about fetishizing and framing Darcy and offering him up to the female gaze' (2001: 112). Hopkins goes on to say that, more broadly, women in the Austen television adaptations 'are given every opportunity and encouragement to exercise the pleasures of looking (ibid.: 120). A rather old-fashioned word can be used to describe the congruence of the gazes of the female protagonist and the female viewer here: this is a brazen gaze, an unashamed looking from a position which, for the character on the screen, is often concealed (she looks hard at the object of her desire from some undetected place in the diegesis) and which for the female viewer is always concealed (from her position on the 'other side' of the screen). Television's brazen gaze is not only associated with this sub-genre of the television serial drama; think of, for example, Milly Buonnano's description of the pleasure of closely and unashamedly scrutinising 'the faces and bodies of political and television personalities, protagonists in stories and events, ordinary people' (2008: 40) discussed in the previous chapter. It does, however, seem to be both implied by the presentation of bodies on screen in the costume drama, and embodied, textually, by desiring figures in the drama.

In the 2015 adaptation of the first of Winston Graham's popular *Poldark* novels (1945–2002), this brazen gaze was once again articulated in the context of a serialised costume drama set in the late eighteenth century. Extra-textually, much was made of Aidan Turner's role as Ross Poldark and his position as a television 'sex symbol' before, during and after the drama was broadcast. The BBC actively marketed the serial on this basis, just as they had done previously with Robin Ellis in their earlier adaptation of these texts (*Poldark*, BBC1, 1975–7).[26] Indeed, before the 2015 adaptation had even begun, the BBC had tweeted a picture of a shirtless Turner having makeup applied to his naked chest in preparation for a scene in which he was scything in a field, a move which had angered its eponymous star: Neelah Debnath in the *Independent* reported:

> [Turner's] apparent disgruntlement at the release of a topless production photograph which was posted on Twitter by the BBC. He told *Newsbeat* at the time: 'I don't know why the BBC are releasing photos of it. It's a bit strange. It's not a stripper show.'
>
> (2015)

Once the series had begun, another marketing strategy, in which viewers were invited to tweet questions to Turner following the fourth episode to form a fan-led interview, descended into farce as Turner was swamped by proposals, compliments and adoration of his physical features. Hannah Furness in the *Telegraph* reported that

> Using the hashtag #AskPoldark, the BBC invited fans from around the world to ask him questions about the show, his career and perhaps the fine art of acting. Instead, he found himself swamped with lustful fans, who were more interested in his hair, chest and sex life.
>
> (2015)

Twitter's role in this account is interesting as it provides a voice to the brazen gaze of the desirous female viewer. Whilst this gaze is brazen precisely because it is concealed and unreturnable, through this social media platform it (rather joyfully) erupts even more brazenly in a collective acknowledgement of television drama's appeal to the female (desiring) gaze. Clearly then something about *Poldark* appealed to the collective female gaze: strikingly, just under 10 per cent of all straight female respondents to the Television Desire survey, which was circulated whilst the series was being broadcast, mentioned Turner as an object of desire.

So why did the fetishisation of Turner's body become, briefly, such a focus for collective female desire in the spring of 2015? Arguably, beyond canny marketing by the BBC, this is largely down to the presence of a desiring female character on screen, a diegetic stand-in for the viewer's brazen gaze and someone who underscores Turner's image as intentional erotic spectacle. Just as an on-screen presenter marks out 'intentional landscape' (as discussed in Chapter 5), here a figure within the diegesis shows the viewers what they should be looking at and how. In *Poldark*, this role was largely taken up by the figure of Poldark's (future) wife, Demelza (Eleanor

Tomlinson), particularly in the first half of the series as their desire for each other develops and results in their marriage, though other women are also seen regarding him as the object of erotic spectacle.[27] Demelza is an unlikely bride – Ross Poldark rescues her from destitution and she initially becomes a servant in his house – and so her gaze on him is initially quite covert, an enforced position in relation to their different social statuses. A good example of this comes in episode two of the first season of *Poldark* where Demelza spies on Ross from the cliff as he swims naked in the sea. This moment of heightened erotic spectacle, which is structured around a shot-reverse shot sequence and accompanied by the romantic theme music of the serial (a Cornish folk-inspired tune played on string instruments), in which bearer and object of the gaze are established, begins as Demelza spots him descending the path to the beach. Her 'concealed' brazen gaze is confirmed by the fact that initially Ross' image is obscured by the grasses she looks through to view him, carefully marking this out as Demelza's point of view. Following a cut back to her face in profile, gazing open mouthed at her employer, a tracking point of view shot up the beach takes in his abandoned clothes, and on into the sea until it meets the sight of his naked back appearing above the waves. A long shot of this same image follows, which Demelza leans into to confirm her position as bearer of the look, followed by a closer shot of Ross bending into a crystal clear sea, his buttocks barely concealed by the angle of the shot (see Fig. 7.2). A profile close-up of her staring intently is then coupled with a medium long shot of him swimming naked. Clearly here, then, we are invited to brazenly gaze alongside Demelza, to be caught in a moment of erotic contemplation in which editing and varying focal lengths serve to articulate her (and the viewer's) desire for a voiceless male body. The voicelessness is particularly marked by the dropping out of diegetic sound in favour of the programme's romantic leitmotif; as in other moments, such as the topless scything scene from episode three referred to above, Demelza's gaze on Ross is accompanied only by the sound of music which offers an aural expression of the emotions she experiences in looking at him. In the scything scene, we initially believe that there is no framing gaze as Ross discusses the fate of Jim Carter (Alexander Arnold) with his servant Jud (Philip Davis) whilst scything in the field; however, once Demelza's gaze on them is confirmed, the sound of their dialogue drops away and once again the same music is heard to

Fig. 7.2 The object of the brazen gaze in *Poldark* (BBC1, 2015–)

connote her desire. Here, then, music is married to the female gaze, and elsewhere in the series – as in the moment when Ross and Elizabeth (Heida Reed) dance in episode two, or Demelza's view of Ross on the cliff again in episode four – it also plays over the top of slow motion sequences in which a desiring gaze is presented. It is almost impossible not to think of a reversal of Mulvey's account of cinematic to-be-looked-at-ness in these moments: 'Man can live out his phantasies and obsessions […] by imposing them on the silent image of woman still tied to her place as bearer of meaning, not maker of meaning' (175: 7). Here the silent image is of man. If, as Mulvey's work has argued, the presence of the woman on screen tends to 'freeze the flow of action in moments of erotic contemplation' (ibid.: 11), in these repeated moments of dialogue-less slow motion, erotic contemplation of Ross Poldark as the central figure of desire in the series is shown.

These moments, and others like them, were reflected on in the Television Desire Survey in the respondents' expression of a preference for the slow revelation of deferred desire and non-explicit representation of sex. In the survey a significant number (just under a quarter) of gay and straight women argued that they found the 'looks and sighs' of a slow building desire in television drama more arousing than watching explicit sex: 'I enjoy watching fiction with sexual tension or attraction. I am not that interested in seeing what actually happens';[28] 'I find the expression of hidden emotion – of emotion that in everyday life has to be expressed – very arousing';[29] 'I often find the build up, the eye sex [...] a lot more erotic than the eventual act.'[30] The respondent who made this final point also argued that 'TV sex is often much better in its gif format on sites such as Tumblr where a fleeting moment, often easily missed, is captured and looped endlessly'. This point focuses the idea of the spectacle of desire not on explicit sex and nudity but in the power of a glance or a minute facial moment; this is an argument that is also made by Charlotte Stevens (2015a; 2015b) in her analysis of the ways that the fanvid reveals the scopophilic looking relations of television. What is key to all of the above examples, then, is the power of the image of a desiring gaze, an image of such power that for many female television viewers it seems even greater than the erotic impact of the object of that gaze. For serial television, with its slow revelation of desire over scenes, episodes or even seasons, this is absolutely key to understanding the erotics of television.

Outlander is another contemporary serial drama which has firmly established a desiring woman at its centre; indeed, blogger and author Jenny Trout called it 'a drama crafted for the straight female gaze' (2014), whilst journalist Maureen Ryan proposed it 'has blown up a lot of the received ideas about sex on television – how it's shot, who it's for, who it's made by and who it's about' (2014b). A British-American co-production based on the historical time travel *Outlander* novels by Diana Gabaldon, the series follows the story of married World War II nurse Claire Beauchamp-Randall (Catriona Balfe) who finds herself transported back to Scotland in 1743 via some magical standing stones, where she encounters rebellion and falls in love with, and marries, the Highland warrior Jamie Fraser (Sam Heughan). From the very outset of this drama, Claire is figured as a desiring and sexually active woman; we see her as a key proponent of the brazen gaze as she regards Jamie, from her appreciation of his muscular physique when tending his wounds in the first episode, to an episode-long contemplation of his body (amongst other things) in the episode 'The Wedding'. Whilst he is a virgin and initially unsure of his role in their sexual relationship despite his physical prowess in other areas (fighting, riding, etc.), she is an experienced woman who guides their love-making; in the latter episode, their approach to each other prior to them consummating their marriage for the second time (out of desire this time, rather than through duty as they had done earlier in the episode) is initially presented in a symmetrical shot-reverse shot in which they longingly regard each other. Here, it is Jamie who removes his clothes first, and Claire who moves towards him to circle his body, guiding the camera around with her. As Jamie stands still, Claire caresses his buttocks until he responds: 'Well then, fair's fair. Take off yours as well.' Here, their desire for each other is slowly drawn out. As they stand naked before each other, we are encouraged through symmetrical composition to again find the erotic in both in their brazen gaze (on each other) as much as on the objects of this gaze (see Fig. 7.3). As Claire asks 'Have you never seen a naked woman before?', Jamie responds, 'Aye, but not one so close.' Whilst this line speaks of his nerves and inexperience, it also offers a commentary on the sequence, and the series at large, in its attempt to bring female desire 'up close' as never before. Gemma Goodman and Rachel Moseley offer an analysis of *Outlander* and *Poldark* (2015) which focuses on the significance of the to-be-looked-at male body, but rather than reading this image in relation to the pleasures of scopophilia alone, they propose that the male body must be read as *territory* in these dramas of national/regional conflict: 'Across [the] bodies [of Jamie and Ross Poldark], we can read the playing out of the politics of region and nation in a period of regional and national instability' (2015). Thus as with the earlier reading of the scenes from *Game of Thrones*, it is possible to see the scopophilic pleasures associated with a gaze on the male body alongside a more richly symbolic handling of intentional erotic spectacle.

Outlander certainly seems to respond almost directly to Jane Juffer's complaint in 1998 that premium cable television fails to fully represent female desire:

> the lack of clitoral emphasis within a context of women's pleasures leaves many of the programs implicitly endorsing a kind of conventional Hollywood intercourse, with the woman inevitably simulating orgasm, issuing the requisite moans and groans as proof of climax, shortly after the man enters her.
> (1998: 203)

Juffer argues, for example, that the 'clitoris does not reign supreme on cable […] [and] cunnilingus is only suggested, and even that suggestion is quite rare' (ibid.: 202). On the contrary, however, *Outlander* features frequent

Fig. 7.3 Symmetrical composition and the desirable/desiring body in *Outlander* (Starz, 2014–)

depictions of Claire being clitorally stimulated by one of her husbands: the series announces this focus from the outset in an early scene when she and her 1940s husband, Frank (Tobias Menzies), explore a highland ruin and he performs cunnilingus on her on a table in the room which will become her place of work in the castle when she is cast back into the eighteenth century. Therefore, whilst the representation of sex and nudity is not entirely unproblematic in this series (there are far fewer shots of male genitalia than those of female full frontal nudity, and the programme's representation of gay male sexuality as predatory and/or sadistic is deeply problematic), *Outlander* does indeed seem to be offering a stark image of women as both the bearers of a desiring gaze and the object of intentional erotic spectacle.

Intentional vs accidental erotic spectacle – flows of televisual desire

This chapter has been focused almost entirely on the question of intentional erotic spectacle, and programming explicitly constructed and marketed as 'erotic' or as appealing to a desiring gaze. However, as suggested in the introduction, this is not the end of the story. In relation to television, erotic sounds and images are part of a much broader flow of programming; the moments focused on in this analysis are fleeting moments within each episode, and every episode is part of a schedule of other programmes in which the erotic might be absent, or in which the erotic might appear quite differently. It is possible to construct a 'viewing strip' (Newcomb and Hirsch, 1994) around a 'hunt' for the erotic across channels and platforms, but more usually viewers stumble across it within the context of their regular viewing. In turn, this flow of programming which is punctuated by erotic spectacle (whether watched as broadcast, or, more regularly now, selected and constructed out of a range of possible sites and platforms) fits into the wider flows of everyday life, and the comings and goings of viewers (families, friends, lovers) around a series of different screens. Therefore, television broadcast and television viewing produce what Barthes termed 'erotic intermittence': 'it is intermittence, as psychoanalysis has so rightly stated, which is erotic: the intermittence of skin flashing between two articles of clothing [...] it is this flash itself which seduces, or rather: the staging of an appearance-as-disappearance' (1975: 10). In the context of television viewing, erotic intermittence, and the appearance and disappearance of erotic spectacle, are fundamental to the structures of television narrative and television scheduling.

Responses to the Television Desire Survey reflected repeatedly on the sometimes uncomfortable intrusion of intermittent erotic spectacle into the flows of programming and everyday life. For example, 57 per cent of respondents found television sex awkward, sometimes excruciatingly so, when shared with others, and argued that it disrupted their engagement with the erotic: 'When watching with other people, particularly family members, I'm more likely to suppress any feelings of arousal.'[31] Three-quarters of these responses were specifically about family members, and it would therefore seem that the idea of the family unit as central to the television viewing experience is upset by the spectacle of sex on TV. Watching, or even stumbling across, the spectacle of eroticism with parents was marked out as particularly difficult by people from all walks of life: 'With my mother it's dreadful as she will say "no thanks, too silly" and undermine the whole thing. Or "Well, there was no need to show that"';[32] 'I'd rather die [than watch sex on TV with my parents].'[33] This relates directly to what Barrie Gunther calls 'the embarrassment caused by bringing out into the open matters that are normally regarded as private' (2002: 12). This discomfort in being seen to view the erotic in public perhaps recalls the clenched jaw and fixed gaze of Michael Sheen as William Masters in *Masters of Sex*, as discussed above. Furthermore, just over a quarter of respondents discussed the fact that they enjoyed watching TV sex more when they are alone, thus confirming a sense of intimacy, discussed above, associated with erotic television. The responses also showed that sex on television interacts with the flows of daily life in a number of ways: people discussed a preference for night time viewing, that sex on TV might make them either leave the room and make a cup of tea, or that it might lead to engaging in sexual activity themselves, and also, interestingly, that the rhythms of the female body (periods, ovulation) had an impact on their engagement with sex on TV.

So viewers' encounters with the erotic via television are intermittent, although the presence of intentional erotic spectacle on television is becoming more common and frequent in various parts of the schedule. However, as Collie's work has shown, erotic spectacle is also found in unexpected places on television. For Collie, this point is tied to the moments where women take visual pleasure in the text which is not anticipated by the programme-maker; she discusses, for example, the women in her study taking scopophilic visual pleasure in the image of male sports stars or actors when being forced to watch sport or Westerns with male relatives, and argues that 'the question of what happens when the male body in programming which has not been specifically imagined to appeal to women, but has nonetheless been objectified by a female viewer, has largely been neglected' (2014: 162). Accounts of this kind of erotic engagement with television, which we might term accidental erotic spectacle, were given throughout the responses to the Television Desire Survey. The young man's desire for the *Gladiators* star discussed above is a good example of this, but women and men also talked about their attraction to television presenters or characters who were not necessarily positioned as the object of a scopophilic gaze in the text: 'I am obsessed with Monty Don';[34] 'I find Rachel Riley on *Countdown* [Channel 4, 1982–] arousing';[35] 'when I was a child I had a serious crush on Scott Tracey from *Thunderbirds*' [ITC, 1965–6];[36] 'I was attracted to the gorillas from the *Planet of the Apes* [CBS, 1974] TV series.'[37] Melanie Williams also confirms the existence of accidental erotic spectacle in her discussion of the sexualised discourse that circulates around the children's television characters Mr Bloom and Mr

Tumble on the UK-based website Mumsnet (2015). In the Television Desire Survey, moments in 'ordinary' television (for example, in soap opera), where the erotic appeal of a scene provided unexpected visual pleasure were also discussed: 'There was an episode of *Coronation Street* in the early 80s where Nigel Pivarro and Michael Le Vell spent a fair bit of time in their briefs as they prepared for a wedding. It was burned into my erotic mainframe!'[38] Whilst these examples come from the testimony of both male and female, and gay and straight, respondents, this latter comment from a gay man encountering accidental erotic spectacle via television at a pivotal moment in the development of his sexual life seems particularly poignant. Indeed, in *Cucumber*, Russell T. Davies dramatises a discussion of a similar moment in the life of his central character, Henry (Vincent Franklin), when, in episode three, his sister (Julie Hesmondhalgh) says,

> When he was about twelve or thirteen, if there was a man on telly who took his shirt off, and believe me, men did take their shirts off back then, Henry would go so quiet, and I mean silent, like he was terrified. Who was that man? Kevin Banks?

At this moment, Henry's face registers his sense of discomfort, pain even, in recalling this encounter with 'accidental' erotic spectacle and responds 'Kevin Banks on *Crossroads* […] Did I look scared?' Here then, and in the accounts of the respondents to the Television Desire Survey's erotic attachments to television, a desirous gaze on an erotic image is not confined to those moments explicitly marked out as intentional erotic spectacle, or to the representation of sex on TV. Rather, the visual pleasures of television are widespread and diverse, subjectively defined according to personal tastes and preferences, and present throughout the medium's broadcast flows. The viewer may seek out (and buy, or download) televisual texts which have been carefully crafted to appeal to a particular set of desires but they might equally stumble upon the spectacle of the unexpectedly arousing via TV. The power of the televisual image to enthral and engage is thus attested to in this discussion of the erotics of television; here, the glance (sometimes imperceptibly) becomes the enraptured gaze.

Conclusion: Sites of wonder, sights of wonder

At 9 pm (BST) on 27 July 2012, the global broadcast of *London 2012 Olympic Opening Ceremony: Isles of Wonder* marked the opening of the London 2012 Olympic Games. The opening ceremony was overseen and conceived by renowned British film, theatre and television director-producer Danny Boyle, and written by Boyle and Frank Cottrell Boyce. The broadcast of the ceremony was a BBC production in collaboration with Olympic Broadcasting Services (which produced the coverage of the 'ceremonial' sections of the proceedings) and Done and Dusted, an independent production company specialising in the filming of large-scale live events. The latter company, under the guidance of director Hamish Hamilton and producer Melanie Fletcher, collaborated closely with Boyle on conveying his vision for the cultural elements of the ceremony. The ceremony was watched by an estimated global audience of 900 million (Ormsby, 2012), and was largely received as both a successful event and television programme that drew on complex ideas about British culture, history and identity to create what Biressi and Nunn term a 'global media spectacle' (2013: 113). Leading up to its broadcast, a great deal had been made of the need for Boyle, Hamilton and their collaborators to 'live up to' the grand spectacle of the opening of the Beijing Olympics five years earlier.[1] However, rather than trying to out-do the grandiosity of that event and broadcast, Boyle focused on producing a televisual spectacle that acknowledged British eccentricity, creativity and ingenuity, that celebrated human endeavour[2] and that paid tribute to those who had had an impact on British social and political life.[3] This was, in essence, the ultimate 'shiny floor' show, a slickly produced and intelligent event and television programme which, as with other examples of the television spectacular discussed at the beginning of this book, combined televisual performance styles and aesthetics with those of other cultural forms (here film, theatre, dance, rock and pop live performance, music video, the catwalk, puppetry, circus, ballet and so on). In its explicit and intricately woven references to diverse aspects of British popular culture, this broadcast and this event thus situated television, alongside other media, as a medium of spectacle and visual pleasure.

Isles of Wonder is a prime example of 'event television' and demonstrates the centrality of spectacle to this genre of programming. In many ways, this programme carefully performs the spectacular pageantry of the television event as explored in Cardiff and Scannell's (1987) analysis of programming of national unity. Furthermore, it can be explored through the critical optic of Dayan and Katz's (1992) delineation of event television in their seminal study of this genre. These theorists examine the relationship between the event depicted on screen and the viewer at home, arguing that television works hard to 'deal with' spectacle (which they equate primarily with the cinema, sports and theatre) and which, they argue, in its purest form places the audience in a passive position, creating a distance and distinction between the spectacle and the audience. They draw on John MacAloon's (1984) suggestion that spectacle in the Olympic Games as event (rather than as television) is 'concerned with grandness, with overviewing the event as a whole, as a display of the combined forces of performers, people and power […] seen from a distance' (Dayan and Katz, 1992: 142). Here, spectacle is a totalising vision which places the audience in its thrall; MacAloon highlights the fact that the word 'spectacle' became taboo in the official vocabulary of the International Olympics Committee in the mid-1980s 'for fear that bigness and dazzle would further emphasize the show-business aspect of the games at the expense of the warmth, magic, and truth which were the reward of genuine involvement [in the Games]' (ibid.: 143). However, Dayan and Katz subsequently argue that event television tries to offer a 'better than spectacular view' (ibid.: 93–4), and they propose that the three-dimensionality of television seeks to provide the viewer with a sense of involvement or immersion, and a privileged access to the event, defying the strict impassiveness of the spectacle/audience relation. They are therefore interested in how domestic viewers relate to the mediatised event (such as an Olympic Games opening ceremony) in 'festive' and 'ceremonial' ways which offer the impression of participation, even whilst they are presented with the grand scale and gloss of spectacle that sets them at a distance from the event depicted on screen.

Whilst citing these analyses of event television acknowledges that this genre of programming is essentially spectacular, and thus offers a further example of spectacular television for this study, this conclusion engages with *Isles of Wonder* in more detail in order to both catalogue and extend the forms of televisual spectacle as explored in this book. An analysis of particular moments from *Isles of Wonder* thus enables a summary of key ideas explored in this work, as well as opportunity for the identification of some gaps in this study. The eclectic nature of both event and programme, which drew on a variety of cultural forms and televisual genres and led to reviewers pronouncing the coverage 'kaleidoscopic' (Anon., 2012), 'complex [...] confused [...] a jolly wonderful mess' (Stanley, 2012), and 'a wild jumble' (Lyall, 2012), offers a diverse array of televisual spectacles and forms of televisual pleasure with which to engage.

As was noted in the exploration of mega-screen television at the mid-twentieth-century exhibition and beyond in Chapter 1 of this book, the 2012 Olympics more broadly was a high point in the communal viewing of spectacular television around large-scale civic screens. As Janet McCabe reported at the end of the Games,

> London was full of big screens and communal TV watching throughout the two weeks of the Olympics. Television turned our international city into a global media village. The Games had us huddling around big screens across the capital, outside stations and in the parks, waving flags – and we did it together [...] People left their homes and tellies to join in these live events, and were never far away from a big screen, as if to prove that we were somehow there. Watching on the big screens translated into a sense of community.
>
> (2012)

Whilst not all television translates easily to the big screen, as discussed in Chapter 1, programming which centres on spectacle and visual pleasure, and which focuses on the ceremonial (as outlined by Dayan and Katz (1992)), clearly finds an easy home in this context. On the night of 27 July, for example, the University of Warwick's open-air 'Big Screen' in the Piazza, one of the central communal areas of campus, was packed with viewers who bore witness to the forms of spectacle produced at the event and in the broadcast (see Fig. C.1);[4] the curators of the screen noted that 'by 8pm, the Piazza steps and bar were full. By 9pm, benches, chairs and rugs had filled the Piazza and the volume had to be increased. The atmosphere was incredible and the party carried on after the ceremony had closed.'[5] Countless similar images of the viewing of this broadcast from around the world showed the same thing: that, as this book has shown, the scale and the formal organisation of spectacular television offer themselves to alternative, large-scale forms of television viewing outside of the home. Whilst Chapter 1 of this book explored historically specific examples of the spectacular public display of television, these public mega-screens can be understood as the most significant contemporary sites of spectacular television.

Isles of Wonder opens not in the stadium but in a frenetic televisual sequence in which a camera (in fact a series of cameras) rushes towards the opening of the games in Stratford, East London. This rapidly paced sequence, following a countdown of 'found numbers' from 50, offers a catalogue of the visual pleasures of television explored in this book. For example, the first shot after the countdown is a composite underwater shot of frogs swimming up from a sunlit river bed, digitally animated and enhanced to give the impression of the close-up view of a beautiful, unseen and unspoilt part of the English countryside (the source of the river Thames, along which the camera will travel). Here, and in the sequence that follows, in which a mobile camera pursues a CGI'd dragonfly along the path of the infant river, *Isles of Wonder* makes explicit reference to the emphasis placed on the beautiful in nature programming more broadly, as discussed in Part II of this study. It thus draws on an established form of spectacular television in its opening moments in order to marshal the viewer's attention. As the camera tracks up and over the trees, a seamless edit is made to an aerial shot which rushes along

Conclusion

Fig. C.1 Big-screen television viewing at the University of Warwick

the course of the river and over a verdant landscape. This landscape and others, such as the sequences of singing children on the stones at Giant's Causeway, in the grounds of Edinburgh Castle and on the vast expanse of Rhossili Beach which shortly follow, as well as the 'landscape' of Glastonbury Tor, which is recreated inside the stadium, form a set of focal images in the opening ceremony which again draw on television-specific visual pleasures outlined in Part I of this book. As with the landscape programmes discussed in Chapter 5, for example, aerial photography features heavily throughout this sequence and later in the ceremony itself; the sense of scale this type of cinematography offers, and the omniscience which is therefore implied, emphasise the breadth and depth of televisual spectacle, and here offer the viewers a set of intentional landscapes which are tied into, and contrasted with, the other spectacular pleasures of the enormous stage set which is constructed in the stadium. These very early landscape sequences in the programme are also shot through filters and digitally enhanced to produce a vividly colourful image; whilst they are not as brightly lit or gaudily colourful as the later montages of fireworks which close the ceremony, the importance of vivid colour in producing arresting televisual images which will grab and hold the viewer from the outset (as discussed in Chapter 2 of this study) is established early on in this sequence.

As the camera makes its progress down the river Thames, it flies over collections of people who stop to look up and wave enthusiastically at it. Whilst these moments are carefully choreographed (there is no suggestion that they have merely noticed a passing camera and turned to wave), they mirror the unrehearsed smiling and waving which punctuates this entire broadcast, and thus the acknowledgement of the moments when the body itself becomes spectacular. From footage of the exuberant actions of athletes suddenly recognising that they are appearing on screen via the video feeds in the stadium, to the shots of spectators in the stadium doing the same, to footage of the thousands of non-professional performers involved in the ceremony greeting a mobile camera in the arena, *Isles of Wonder* is full of bodies caught up in the awareness of *becoming spectacle*, of the moment in which one *makes a spectacle of oneself*. As with the crowds clamouring to 'see yourself on television' described in Chapter 1 of this book, or many of the spectacular bodies analysed in Chapter 6, these are voiceless bodies, bodies that often only for a second or two become absorbed in the televisual spectacle. The shift between the grand scale of the aerial view and the close-up on the individual faces of performers and participants thus affords the viewer of *Isles of Wonder* two very different, but recognisable, images of televisual spectacle as explored in this book.

Whilst it is not possible to look in detail at every aspect of the ceremony broadcast which might be viewed as harnessing the spectacular in this conclusion, the sequence 'Frankie and June say… thanks Tim', which acknowledges the impact that Tim Berners-Lee's invention of the World Wide Web has had on British social life, is particularly significant. The sequence, which shows a young couple preparing for and enjoying a night on the town where they meet, flirt (via text messages) and then fall in love, interweaves reference to a myriad of popular music, film and television to create a love story for the 'digital age'. Following the opening of this sequence, where scenes of a young woman from an 'ordinary' family are intercut with a montage of television programmes and films that tell the story of her preparing for her night out, she joins a group of dancers on stage. In this sequence, we see two groups of friends gradually come towards each other, and their use of social media is tracked in text and image boxes which appear onscreen. As the young couple regard each other or look for each other amongst the chaos of frenetic dancing and flashing lights, the television cameras pick their desiring looks and smiles out of the crowd, offering a surprisingly intimate story of desire in the context of an Olympics opening ceremony. The fact that their flirtatious texts are represented on screen adds to the intimacy of this story within the ceremony, referencing earlier televisual uses of this conceit (particularly in UK teen soap opera *Hollyoaks*, Channel 4/E4, 1995–, as an early developer of this technique from 2010 onwards), and building an intimate picture of their desire. As the young couple embrace and kiss for the first time, they do so in front of a 'house' which is actually a large three-dimensional screen in the centre of the stage, onto which is being projected a series of images of romantic clinches made throughout the history of the moving image. Whilst most of these interpolated images are of film couplings, one of the last and most significant images we see on the house-screen is the 1994 kiss of Beth Jordache (Anna Friel) and Margaret Clemence (Nicola Stephenson) in UK soap opera *Brookside* (Channel 4, 1982–2003), the first pre-watershed kiss between two women on British television.[6] This sequence, which references the depiction of romance in recent teen soap opera, thus acknowledges the centrality of desire to television as well as film. The intimacy of the looks and smiles of the flirting couple, and the invitation to the viewer to participate in these structures of looking, can thus be read through the exploration of the erotics of television as outlined in Chapter 7 of this study, or in relation to Janet McCabe's suggestion that, in its reference to a myriad of television programmes and genres, *Isles of Wonder* was truly a 'spectacle made for the small screen' (2012).

As well as interpolating the films and television programmes discussed above, this section of the ceremony/broadcast is also interesting for the ways it incorporates music video, also via the house/3D screen as backdrop. A frenetic cycle of videos conveying the excitement of a night on the town via reference to British pop music, including Queen's 'Bohemian Rhapsody', Soul II Soul's 'Back to Life' and The Prodigy's 'Firestarter', is intercut here with a chronologically evolving cycle of British films and television programmes. I point this out as it highlights one of the significant absences in this study, as do the earlier and later interpolations of sports programming via montage sequences which recall past sporting achievements through the medium of television in *Isles of Wonder*. Both the sports programme and the music video must be seen as key examples of spectacular television; importantly, they are both genres of programming which frequently incorporate montages of spectacular images, confirming my argument that montage is to television spectacle what the long take and the wide angle shot are to filmic spectacle. Whilst the sports programme often offers what I describe in Chapter 5 as 'narrative montage', that is, montage as form of 'recap' which tells us what has just happened in the unfolding story of a match, race or bout, both sports television and the music video also frequently operate around what we now understand as spectacular montage, and offer moments (particularly in the closing sections of sports broadcasting) where we are invited to simply look and enjoy, to sit back and relax, to contemplate a series of beautiful or striking images on screen. The broader applicability of the term 'spectacular montage' outside of the genres it is applied to in this study is thus acknowledged here.

We see this very obviously, for example, in the closing of this opening ceremony broadcast, following the lighting of the Olympic flame cauldron. Over the sound of Pink Floyd's 'Eclipse', increasingly wide angle aerial shots of the Olympic stadium, and shots from inside the stadium looking up to the fireworks which shower overhead, are intercut with 'classic' televisual images of international sporting triumphs. Here then we see a montage of images of two different types of pro-filmic spectacle: images of a site rendered spectacular as part of an event (the opening ceremony of the 2012 Olympic Games) and images from sporting events throughout the history of television. Whilst this ceremony and the races, events and matches depicted here can be seen as spectacular in and of themselves, their presentation on television intensifies, rather than dilutes, our understanding of them as spectacle. Furthermore, this montage, which brings both the sites and the sights of television spectacle together, structures these images into the spectacular montage in order to invite the viewer to sit back and watch, to marvel, to enjoy: in short, to take visual pleasure from them.

In closing it must be pointed out that whilst I have shown the Olympic Opening Ceremony to be another prominent and obvious instance of intentional television spectacle which is structured around the visual pleasures of the medium in particular ways, the 'event' status of this broadcast, its prominence on large communal screens around the UK and beyond, and

Conclusion

thus its separation from 'regular' television must be addressed. As has been seen in each chapter of this study, spectacle is inherent within and across a myriad of televisual genres and throughout the medium's history; it is not reserved for 'special' or 'ceremonial' programming, but rather is part of the day-to-day fabric of television broadcasting. Spectacle is often found in unusual places on television, and the medium's visual pleasures are frequently as important to its viewers as the unfolding of narrative or the conveyance of information. Spectacular television, then, offers the viewer moments in which they contemplate beauty, the erotic, the grotesque and the breathtaking; it presumes a relationship between spectator and programme which is intense rather than casual, and which is punctuated by moments of powerful visual pleasure.

Notes

Introduction

1 See http://www.oxforddictionaries.com/definition/english/spectacle, accessed 13 December, 2015.
2 See for example my discussion of the applicability of screen theory to a reading of the erotics of television in Chapter 7.
3 See Terrace (1995: ix) for a discussion of NBC's colour 'spectaculars' in the 1950s.
4 Warhol's films *Sleep* (US, 1963) and *Empire* (US, 1964), in which he experimented with filmmaking consisting of long takes of his sleeping friend John Giorno (for 5 hours and 20 minutes) and the Empire State Building (for 8 hours) respectively can be seen as precursors to this form of programme making.
5 According to the BBC4 website: http://www.bbc.co.uk/programmes/b05ttkx2, accessed 13 December, 2015.
6 Twitter and Facebook were used to circulate surveys for Chapters 4 and 7.

1 Television comes to town

1 This was the annual North London Exhibition, a kind of local Ideal Home show, with additional fashion sections.
2 See Bird Jr (1999) on television at New York's World's Fair, for example.
3 According to the shooting script held in the following file: BBC WAC, T14/436/3 TV Outside Broadcasts, Festival of Britain, South Bank, File 3A, 1951.

4 Undated/unsigned memo in the following file: BBC WAC, T14/436/6 TV Outside Broadcasts, Festival of Britain, South Bank, File 4B, 1951.
5 The programme had an average shot length of 22 seconds, and the shooting script documents a long series of slow (30+ second) pans from, for example, the Shot Tower to the Unicorn or from the Dome of Discovery to ships on the Thames. The longest of these was a 55-second pan from the Dome to Charing Cross and the Shell-Mex building.
6 As detailed in the musical rights contract for the programme held in the Outside Broadcast Unit's Festival of Britain, Southbank file BBC WAC, T14/436/6.
7 N.B. The title 'Telekinema' was used on the side of the building at the Festival and in most of the subsequent coverage of this exhibit. 'Telecinema' was used in the BFI's planning documents and in the Festival programme, however.
8 This technology was invented by Baker Smith as a scenic design technique for theatre and television drama. It was first used on television in a production of *Julius Cesar* on 24 July 1938, and was described in the *Radio Times* thus: 'Briefly, it consists of an arrangement of 2-kilowat spot lamps, by which shadows and part shadows can be cast upon a translucent screen [...] One effect can be changed to another without any movement other than the touch of a switch' (The Scanner, 1938).
9 Unfortunately this is just out of shot on Fig. 1.3, as it was placed in the entrance way to the Television pavilion.
10 Designed for the Festival by Edward Mills.
11 Sarah Easen (2003) offers a detailed account of the films that were both planned for, and actually shown at, the Telekinema, including a discussion of the five documentaries which represented the work of British film industry in the Telekinema. These were *Air Parade* (Bill Napier, 1951), *David* (Paul Dickson, 1951), *Forward a Century* (J. B. Napier-Bell, 1951), *Waters of Time* (Basil Wright, 1951) and Humphrey Jennings' *Family Portrait* (1951).
12 Dennis Forman wrote in *The Times*: 'There is no doubt that the Telecinema has proved one of the most popular features of the [Festival of Britain]. More than a quarter of a million members of the public attended the performances given during the first ten weeks of running, while many others were unable to gain admission'. (1951).
13 Entry fee to the Telekinema was 5 shillings.
14 In the UK, for example, the main public demonstration event for television in 1930 was a fortnight of television broadcasts produced by the Baird Television Company at the Coliseum in London as part of a larger evening of variety performances in July and August. Cinemas, and companies such as Scophony Ltd., continued to experiment with the exhibition of television in the first decades of a regular television service. Beyond the UK, Knut Hickethier has documented the development of cinema-based television under the Third Reich (1990),

whilst Janet Wasko briefly discusses the development of the Theater Television Network in the US, offering the cinematic screening of 'sports events (such as boxing and collegiate games), public affairs, and entertainment events' (2003) from TV.

15 A note is written on the edge of a memo from Admiral Dorling (on behalf of the Radio Industry Council) by A. J. P. Hytch, the Assistant Head of Publicity, that the 'DG does not want entire emphasis on television'e(Dorling, 1946).
16 This was not a new technique. At the 1938 Ideal Home Exhibition the centre of the television exhibit was a 2,000-square-foot studio with three glass sides producing seven closed-circuit broadcasts a day which could be watched on TV screens within the Exhibition or from the outside of the glass-walled studio.
17 With stills and exhibits from panel games and the flashing timepiece from 'Beat the Clock', the game within *Sunday Night at the London Palladium*.
18 Including the studio set from *Emergency Ward 10* (ATV, 1957–67) and back projection showing recent items from *This Week* (A-R, 1956–78).
19 Featuring stills from *The Jack Jackson Show* (ATV, 1955–9) and *Cool for Cats* (A-R, 1956–61), alongside an elaborate jukebox.
20 With an innovative costume display hung inside large 'peep-show recesses in baroque designs' (Anon., 1958).
21 Complete with stuffed bison and a tepee.
22 It was sold by the *Daily Mail* to the events and publishing company Media 10 in 2009.
23 According to Ryan (1997: 17).
24 According to the Ideal Home Exhibition Catalogue, 1962.
25 Other relatively recent forays into variety show territory have been unsuccessful: for example *The Nick and Jessica Variety Hour* (ABC, 2004).
26 See Chapter 4 for a discussion of *The Blue Planet Prom in the Park* (2002).

2 Spectacular colour? Reconsidering the launch of colour television in Britain

1 Briggs (1995) offers more of this, as does Andreas Fickers' account of the struggle for a European-wide colour standard (2010).
2 The British Radio Equipment Manufacturers' Association (BREMA) 1967 film, 'The Colour Television Receiver', which details the process of setting up colour television in the home, can be viewed via YouTube: http://www.youtube.com/watch?v=BkrLf_9n_7w, accessed 13 December, 2015.
3 'Weekly Review of Programme Presentation minutes' 28/11/67, BBC WAC, T5/682/1, *Vanity Fair,* General and Filming.
4 Memo from George Campney (Head of Publicity) to David Attenborough, 'Colour Promotion', 27/6/67, BBC WAC file R44/773/1, *Publicity: Television*

Colour System: 'I gather that most of the critics now have colour receivers – courtesy of *Radio Rentals* so we should not be short of a notice or two.'

5 Incidentally, there is a letter from Buckingham Palace from the middle of December, 1967, in the *Colour Television: General* file at the BBC WAC (R78/554/1) thanking the Corporation for the offer of a colour set but stating that 'The Queen has only just (in the last few days) ordered and had installed a set at Windsor Castle' (13/12/67), so clearly even the Royal Family were initially slow to adopt colour television.

6 Introduction of the BBC Publicity Service dossier, 'Sample of Press Coverage: *Colour Television Comes to Town*', BBC WAC file R78/555/1, *Colour Television Promotion*.

7 Board of Management minutes, point 629g, 22/7/68, BBC WAC file R78/555/1, *Colour Television Promotion*.

8 Board of Management minutes, point 699i, 19/8/68, BBC WAC file R78/555/1, *Colour Television Promotion*.

9 Geoffrey Howard (Exhibition and Campaign Organiser), 'Wimbledon 1968: Colour TV Store Promotions', 18/6/68, BBC WAC file T23/135/1 *TV Publicity: Colour Television 1966–1968*.

10 Ian Atkins 'Working in Colour: A lecture by Ian Atkins, Controller, Programme Services, Television, at the third of the sixth series of lunch-time lecture in the Concert Hall of Broadcasting House' 13 December 1967, BBC WAC file T66/161/1, *Colour TV Start of Services 8/10/53-15/10/71*.

11 Kenneth Lamb 'Colour Television: A Report on Progress' 7/3/68, BBC WAC file R78/1967/1, *Colour Television Policy*.

12 Phil Ward 'Commendation' 1/4/68, BBC WAC file T12/1188/1, *Once More With Felix 16/12/67-30/3/68*.

13 David Attenborough 'Colour Policy', 5/12/66, BBC WAC file T16/47/10, *TV Policy: Colour Television File 7B, Aug-Dec 66*.

14 Made up of selected episodes of the award-winning anthology series *Theater of Stars* (NBC, 1963–67).

15 Barry Learoyd 'Colour Familiarisation Course Number 12' 6/1/67, BBC WAC file T5/1015/1, *Drama: Colour Courses 10/2/67*.

16 Ian Atkins 'Working in Colour: A lecture by Ian Atkins, Controller, Programme Services, Television, at the third of the sixth series of lunch-time lecture in the Concert Hall of Broadcasting House' 13 December 1967, BBC WAC file T66/161/1, *Colour TV –Start of Services 8/10/53-15/10/71*.

17 Kenneth Lamb 'Colour Television: A Report on Progress' 7/3/68, BBC WAC file R78/1967/1, *Colour Television Policy*.

18 David Conroy 'BBC Classic Serials: Colour Programmes, General Notes' 27/10/67, BBC WAC file T5/1809/1, *Portrait of a Lady, General*.

19 See 'Programme Colour Costs' document, which shows the differential costs of producing in colour and black and white across all genres of

programming: 11/11/66, BBC WAC file T39/9/1, *Controller of Television Admin Papers: Colour Television Colour BBC1 1966–69*.
20. Shaun Sutton 'Memo to C. P. Tel.' 12/7/67, BBC WAC file T5/682/2, *Vanity Fair, General and Filming*.
21. Ibid.
22. Indeed, perhaps this is why the novel was also the source for the first feature film to be shot in Technicolor, Robert Mamoulian's *Becky Sharp* (US, 1935).
23. Although shot in colour, *The Owl Service* would be first broadcast in black and white because ITV did not begin colour broadcasting until the end of 1969.
24. James Ferman, 'Memo', BBC WAC file T5/1620/1, *Drama: Lovely In Black (Thirty Minute Theatre) 24/1/68*.
25. David Conroy 'Press Release for initial broadcast of *Vanity Fair*' 6/10/67, BBC WAC file T5/682/1, *Vanity Fair: General and Filming*.

3 At home on safari

1. Paul Gilroy also draws on Pratt's notion of the 'contact zone' in his discussion of the imperial metropolis (2000: 117).
2. This chapter concentrates solely on the Denises' work for the BBC partly because no copies of their series *Michaela and Armand Denis* (ITV, 1955) and *Armand and Michaela Denis* (ITV 1957–8) exist, nor have I been able to find any paper archive in relation to these programmes.
3. Lauren Van Vuuren's 2003 article on Laurens van der Post's *Lost World of the Kalihari* (BBC, 1958) is an exception. Van Vuuren offers a lucid account of Van der Post's six-part series, outlining the ways in which it romanticised, and infantilised, Bushmen culture: she argues that this series offered the viewer a 'journey out of time' and steadfastly refused to show any sign of acculturation amongst the people at its centre. According to Van Vuuren, the *Lost World of the Kalihari* was to become a highly influential documentary in the latter part of the twentieth century, despite protestations from anthropologists and development specialists about its simplistic representation of Bushmen culture.
4. I am particularly indebted here to the telling of Kenya's colonial history in C. S. Nicholls' fascinating book *Red Strangers: The White Tribe of Kenya* (2005).
5. See Sleeman, 1947; MacKenzie, 1988; Steinhart, 2006; and Nicholls, 2005 for a much fuller account of the history of safari in Kenya.
6. Approximately £19,000 when adjusted for inflation (see http://safalra.com/other/historical-uk-inflation-price-conversion/), accessed 13 December 2015.
7. See Sleeman, 1947 for an account of this shift.
8. Wendy Webster (2011) makes a similar argument about the film *Where No Vultures Fly* (1951), which tells the story of Robert Payton's quest to establish a National Park in Kenya.

9 See, for example, the following: Ambler, 2011; Cameron, 1994; Chapman and Cull, 2009; Codell, 2011; Geraghty, 2000; Grieveson and MacCabe, 2011a, 2011b; MacKenzie, 1986b; Maynard, 1974; Rice, 2011a, 2011b; Richards, 1986; Webster, 2011; and the brilliant Colonial Film database (http://www.colonialfilm.org.uk/, accessed 13 December, 2015.).
10 See Grieveson, 2011.
11 See Rice, 2011a.
12 See Smyth, 1979; Rice 2011b; and Rice's commentary on films such as *Better Homes* (1948) at http://www.colonialfilm.org.uk/, accessed 13 December, 2015.
13 They made *Stampede* (British Instructional Films, 1930) and *Struggle for Life* (Foy Productions, 1935).
14 Set in Malaya, not Africa.
15 RKO's most profitable film of 1949, according to Beinart and Coates, 1995.
16 The Denises were relatively unusual in that they made programmes simultaneously for the rival channels, a fact that their producers at the BBC didn't like, thinking that the association with ITV downgraded their programmes for the Corporation.
17 Armand was eighteen years Michaela's senior.
18 Hazel Collie's research for the AHRC-funded project 'A History of Television for Women in Britain, 1947–89' (University of Warwick/ De Montfort University) has attested to the success of this address. As a 66-year-old Liverpudlian woman stated: 'I was absolutely transported. Smitten. And then when I actually managed to get to Kenya, sit in a Jeep and see a cheetah, I burst into tears.'
19 This episode is streamed online as part of the fantastic online history initiative in the field of wildlife film and television history, *Wild Film History*: http://www.wildfilmhistory.org/film/259/Filming+Wild+Animals.html.
20 See for example Spigel, 1988.
21 As explored in Ellis, 2000.
22 Webster's key examples are Catherine Munro in *Men of Two Worlds* (1946), or Mary Crawford in *Simba*.
23 Headlines such as 'Braving the jungle with lipstick and mascara' (Anon., 1955) and 'Leopard-girl says: I'll talk peace to the Mau Mau' (Pugh, 1954) were typical.
24 Nearly one-third of all Kamba men were enlisted in the King's African Rifles, Britain's East African colonial army, in the mid-1940s, compared to only 6 per cent of Kikuyu men (Parsons, 1999: 683).
25 *The Michaels of Africa* (BBC, 1955–6).
26 This idea is extended in the exploration of more recent natural history programming in the following chapter.
27 Nicholas Crocker, Letter to Armand Denis. 25/2/64, BBC Written Archives Centre, Caversham – SW3/20/1.

4 Visual pleasure, natural history television and televisual beauty

1. A similar argument is also made in Chapter 5 in relation to the depiction of landscape on television.
2. This is, of course, partly because much of this series was shot under the sea at extreme depths and it would therefore not be possible for this veteran presenter to have been present.
3. See Chapter 5 for a similar discussion of the screensaver aesthetic.
4. See episode two of *Frozen Planet* for the latter.
5. I respond here to Sarah Cardwell's suggestion that applying theories of philosophical aesthetics to the analysis of television might allow us to better understand concepts such as beauty, harmony and the sublime in relation to the medium (2013: 40).
6. These theories are discussed at length in Armstrong (2005: 26–34).
7. The survey was sited on the Department of Film and Television Studies website at the University of Warwick and was shared via social media (Facebook, Twitter), using a snowballing technique in order to reach 162 respondents through sharing/retweeting. Just under a third of the respondents were male, and the age of respondents ranged from 21 to 71. Ten per cent of respondents were based outside of the UK.
8. This is the fictional equivalent of the contemplative figure in landscape programming discussed in Chapter 5.
9. Quote taken from Scruton's narration for the programme *Why Beauty Matters* (BBC2, tx. 28/11/09).
10. I think it is correct to say that almost every type of programme except news was represented in the list of specific programmes that was produced by this viewer research.
11. This respondent was one of the eight people who responded to the survey saying that they would not describe any television as beautiful.

5 Television's landscapes, (tele)visual pleasure and the imagined elsewhere

1. See Pidduck (1998) for an excellent analysis of landscape in the heritage drama.
2. See Chapter 3 of this book, for a discussion of landscape in the colonial natural history programme.
3. See Wheatley (2004) for a discussion of visual pleasure in the natural history programme (partly reproduced in Chapter 4 of this book) and Caughie (2000: 20–5) and Wheatley (2005) for a discussion of the visual pleasures of the literary adaptation.

4 See for example the company 'Frame Your TV': https://www.frameyourtv.co.uk, accessed 13 December 2015.
5 See for example the company 'FramethatTV.com': http://framethattv.com/motorized-art/.
6 See Holdsworth (2006) for an early discussion of slow television. Slow TV is also a programming format, discussed at greater length in the introduction of this book, which was developing in Germany and Norway contemporaneously with the broadcast of these landscape documentaries.
7 See http://www.bbc.co.uk/archive/aerialjourneys/5316.shtml, accessed 13 December 2015. Several episodes are streamed via the BBC website as of October 2010.
8 The exception to this is the second shot in this sequence, which tracks around the outside of a modern high-rise block of flats in an unidentifiable urban location.
9 For a discussion of the negotiation of national identity in *Coast* see Thompson (2010a).
10 All quotes from Richard Mervyn are taken from a personal interview with the author, 8/9/09.
11 In fact, Richard Mervyn was forward thinking in having already proposed a form of Slow Television, a series of helicopter journeys without narration, which he saw fitting into one of the short 'post-news' broadcasting slots, to various commissioners prior to my interview with him and before the genre's huge popular success on Norwegian public television (see discussion of slow TV in the Introduction of this book).
12 The idea of a Windows-based computer photo montage in television's handing of still images has also been discussed by Jason Jacobs (2009).
13 All quotes from Steve Evanson are taken from a personal interview with the author, 16/10/09.
14 Andrew Higson (2006) makes a similar argument about the representation of rural spaces in British cinema.
15 My thanks to Hannah Andrews for pointing this out to me.
16 The programme ended in 2007.
17 The programme ended in 2003.
18 See Turnock (2007: 147), for an account of this.

6 Fascinating bodies

1 See Perkins (2014) and Shacklock (in prep.) for a discussion of television kinaesthetics.
2 Schools programming on the BBC began in the autumn of 1957.
3 The series was telerecorded and re-shown in 1962 and 1964.
4 My thanks to the BBC archive for enabling me to view a number of episodes of *How Your Body Works*.

5 The series was telerecorded and re-shown in 1970 and 1974, but is unfortunately no longer available to view.
6 This has also been the case in more recent years with the series *Superhuman* (BBC1, 2000) and *How to Build a Human* (BBC2, 2002). It is greatly unfortunate that none of this programming from the 1960s appears to have survived.
7 *Life Before Birth* (BBC, 1960), *Replacements for Life* (BBC, 1960), *The Science of Man* (BBC1, 1964–5), *Dr Alice Roberts: Don't Die Young* (BBC2, 2008) plus occasional episodes of the long-running science current affairs programme *Eye on Research* (BBC, 1957–61) and documentary series *Man Alive* (BBC2, 1965–82).
8 I concentrate here on the presentation of the body's organic structures, and particularly the spectacular presentation of the workings of the internal organs. For an extended exploration of the body at a molecular or genetic level, see the work of Sofia Bull (2012; forthcoming, 2017).
9 'Once it reaches a quiet corner of your lungs, the spore implants, to grow. Left unchecked it would fill this entire chamber [...] Within weeks you'd be dead [...] [Your body then] releases macrophages, prowling, marauding, remorseless killers.'
10 Indeed, the musical idioms of science fiction are frequently used as an aural backdrop to such spectacular landscapes, as in the *Incredible Human Machine* sequence where we go 'into' the brain.
11 In the later episode 'Blood and Vessels' he demonstrates the impact that exercise has on the heart by holding a microphone to his own chest, before and after running around outside the television studio.
12 In *Gunther's ER* (Channel 4, 2007).
13 Susan Sydney-Smith also describes this coupling in relation to the presentation of the female cadaver in the British crime drama (2007: 196).
14 *Bodyshock* regularly draws in some of Channel 4's largest viewer figures; the episode 'Half-Ton Mum' was watched by 4.8 million people, giving the channel an 18 per cent share of the audience, compared to *New Tricks* (2003–) on BBC1, which had a 20 per cent share, and the finale of ITV drama *The Fixer* (2008), which had a smaller 17 per cent share of viewers (Dowell, 2008). Indeed, Stuart Heritage (2013) equates the viewing figures for 'The Man with the Ten Stone Testicles' with those of popular programming such as *The One Show* (BBC1, 2006–) and *Big Brother* (Channel 5, 2011–) to underscore its popularity.
15 Who was exhibited at the Brighton Aquarium in the UK in 1886.
16 Who toured Europe as a sideshow exhibit between 1873 and 1904.
17 This is true of the Channel 5 series *Extraordinary People*. Prior to the broadcast of the episode 'Octopus Man' on Channel 5* on 21 July, 2012, the announcer's voiceover accompanies shots of a fairground (one of 5*'s idents at the time): 'Oh, *Extraordinary People* now, and this time it's the Octopus Man. There's

graphic scenes from the operating table and images which some viewers may find upsetting.'
18 See also the *Bodyshock* documentaries 'Born with Two Heads' (tx. 20/2/06), 'The World's Biggest Boy' (tx. 26/11/08), 'I Am the Elephant Man' (tx. 7/4/08) and 'The Man with the Ten Stone Testicles', for example.
19 See Frances Bonner (2005) and Catherine Belling (1998) for two excellent accounts of the spectacle of surgery on television.
20 For example *Embarrassing Bodies: Kids* (Channel 4, 2010), *Embarrassing Old Bodies* (Channel 4, 2010), *Embarrassing Fat Bodies* (Channel 4, 2010) and *Embarrassing Bodies Down Under* (Channel 4, 2013).
21 *Embarrassing Bodies*' BAFTA award-winning website is responsible for 42 per cent of Channel 4's web traffic' (Wiseman, 2010).
22 See Davies (2011) for a discussion of the public service imperatives of this site.
23 Episode Four of Series Four.
24 James Foley and Steven Sotloff.
25 Hervé Gourdel.
26 David Haines, Alan Henning and Peter Kassig.
27 My thanks to Jake Edwards for bringing this correlation to my attention.
28 The programme included stills of the artist Zhu Yu eating what appears to be a stillborn infant. One shot shows him washing the baby before he eats it and three others depict him biting into the body. Zhu Yu then admits he was sick afterwards.
29 The Tokyo exhibition between 1996 and 1998 had 2.5 million visitors and 1.4 million people visited the Berlin exhibition in just seven months of 2001 (Harris and Connolly, 2002: 9).
30 Von Hagens has also been compared to Dr Frankenstein (Harris and Connolly, 2002: 9), Hannibal Lecter (O'Rorke, 2001: D5) and 'a cross between evil scientist and the television cliché of ghoulish murder scene pathologist' (Searle, 2002: 11).

7 The erotics of television

1 See particularly the episode 'A God in Colchester' (tx. 29/11/76). My thanks to Barry Langford for bringing this back to my attention.
2 Drawing on Terry Castle's (1993) work on this concept.
3 The questions were: (1) Have you ever been attracted to anyone on television? (2) Are you more likely to be attracted to a fictional character, a television presenter, a public figure (e.g. politician, sports personality) or a 'real person' (e.g. reality show/documentary participant) on TV? (3) How do you feel about watching sex on television? Do you enjoy, avoid or feel indifferent about this? (4) Have you ever found a television programme erotic/arousing? If yes,

please give further details. (5) Are there any circumstances in which you are either more or less likely to enjoy watching sex on television? (6) Does what you watch television on (i.e. TV set, laptop, iPad) have an impact on your 'erotic engagement' with television? (7) Have you ever watched porn on your television? If yes, please state how it was screened (i.e. via DVD, video, specialist cable channel, etc.).

4 The largest group who responded were heterosexual women (48), though I also received responses from bisexual women (14), straight men (12), gay men (4) and lesbians (5).
5 Arthurs makes a similar argument about the launch of Channel 5 in the UK, with its emphasis on 'films, fucking and football' (2004: 42).
6 See for example Ryan (2014a); Itzkoff (2014).
7 Allen was already known for being comfortable with nudity, having appeared naked on stage in the revival tour of *Equus* in 2008.
8 This resonates with the feminist scholar Elizabeth Wilson's rejection of psychoanalysis (see Gamman, 1998: 18).
9 See Perkins (2014) for an exploration of the bodies of *Girls* and our kinaesthetic responses to them.
10 This relationship between children and early objects of desire on television was also confirmed in Andy Medhurst's paper on camp in children's television (2015).
11 See also Wheatley (2015) for a discussion of Mulvey's ideas as they relate to television.
12 Stevens (2015a; 2015b) provides an excellent analysis of the ways in which the fanvid might be understood as illuminating the intensity of scopophilic looking relations on television.
13 This figure excluded the 12 per cent of people who had never watched television on anything other than a television set.
14 A 33-year-old straight, female college instructor from the US.
15 A 29-year-old straight, female student from Switzerland.
16 A 44-year-old straight, female academic from the UK.
17 A 41-year-old straight, female artist from the UK.
18 A 28-year-old straight, female librarian from the UK.
19 A bisexual office worker in her thirties from the UK.
20 A 26-year-old bisexual, female buyer from France.
21 A straight, female academic in her forties from the UK.
22 A German straight, female language teacher in her thirties.
23 A lesbian musician in her thirties from the UK.
24 A straight, female academic in her thirties from the UK.
25 A German straight, female language teacher in her thirties.
26 See Moseley (2013) for a discussion of this earlier series and the emphasis on Poldark as romantic hero in it.

27 As in the party scene at the beginning of episode two, where we see a brief montage of women watching Ross at the dance.
28 A Finnish straight, female researcher in her late forties.
29 A straight female HE worker in her forties living in the UK.
30 A lesbian in her forties living in the UK.
31 A 34-year-old straight, female academic from the UK.
32 A self-employed, 40-year-old straight woman living in the UK.
33 A straight, female PR consultant from China in her thirties.
34 A straight female college lecturer in her forties from the UK.
35 A straight male company director in his fifties from the UK.
36 A straight female media officer in her twenties from the UK.
37 A straight female teacher in her forties from the UK.
38 A gay staff nurse in his late forties from the UK.

Conclusion

1 See Gibson (2012), for example.
2 In its reference to the Industrial Revolution, the Tommies of World War I, and the inventor of the world wide web, Tim Berners-Lee, for example.
3 Performers represent the Suffragettes, the Jarrow hunger marchers of 1936 and Caribbean immigrants arriving on the MV *Empire Windrush*, for example.
4 My thanks to Sarah Wall, curator of the University of Warwick's Big Screen, for sharing her account and images of this broadcast with me.
5 From correspondence with Sarah Wall, University of Warwick, 22 October 2012.
6 According to entertainment blog AfterEllen, this moment of *Isles of Wonder* was also the first same-sex kiss seen on television in 76 countries, including Saudi-Arabia (Linster, 2012).

Bibliography

Aaron, M. (2015) *Death and the Moving Image: Ideology, Iconography and!*, Edinburgh: Edinburgh University Press.

Aaronivich, D. (1998) 'TV Review: How Herbie died for all of us', *Independent Online*, 28 June, http://www.independent.co.uk/life-style/tv-review-how-herbie-died-for-all-of-us-1168089.html, accessed 22 December 2014.

Adam, K. (1947) 'Memo', BBC WAC, R44/148, 23 December.

Ali, B. (2011) 'Liquored Up "Wid de Sperrits": Gothic Figures of Black Men in White Boys' Adventure Stories', *Journal of Popular Culture*, 44, 6: 1141–70.

Allen, R. C. (ed.) (1992) *Channels of Discourse, Reassembled*, London and New York: Routledge.

Ambler, C. (2011) 'Projecting the Modern Colonial State: The Mobile Cinema in Kenya', in L. Grieveson and C. MacCabe (eds) *Film and the End of Empire*, pp. 199–223.

Andrejevic, M. (2004) *Reality TV: The Work of Being Watched*, Lanham: Rowman & Littlefield.

Andrew, D. (1984) *Concepts in Film Theory*, Oxford: Oxford University Press.

Andrews, M. (1999) *Landscape and Western Art*, Oxford: Oxford University Press.

Anon. (1937) 'Is Television within Your Reach?', *Ideal Home*, November: 407.

_____(1955) 'Braving the Jungle with Lipstick and Mascara', *Sunday Express*, 22 May: 22.

_____(1958) 'The 1958 Radio Show Supplement', *The Times*, 28 August: 1.

_____(1967a) '40 Hours' Colour in First Week of July', *Electrical and Radio Trading*, 29 June, held in press clippings pack, BBC WAC file R78/555/1.

_____(1967b) 'It's Colourific', *Sunday Mirror*, 2 July, held in press clippings pack, BBC WAC file R78/555/1.

Bibliography

_____(1967c) 'Charlie's Spectacular', *Radio Times*, 7 December: 11.
_____(2012) 'Kaleidoscopic Pageant Sets London Games Rolling', *China Daily*, 27 July, http://europe.chinadaily.com.cn/sports/2012-07/28/content_15625614.htm, accessed 16 October 2015.
Armstrong, J. (2005) *The Secret Power of Beauty*, London: Penguin.
Arroyo, J. (2000) *Action/Spectacle*, London: BFI Publishing.
Arthurs, J. (2004) *Television and Sexuality: Regulation and the Politics of Taste*, Maidenhead: Open University Press.
Aston, J. (2012) 'Television X-cised: Restricted Hardcore and the resisting of the real', in J. Aston, B. Glynn and B. Johnson (eds) *Television, Sex and Society: Analyzing Contemporary Representations*, pp. 79–92.
Aston, J., Glynn, B. and Johnson, B. (eds) (2012a) *Television, Sex and Society: Analyzing Contemporary Representations*, London: Continuum.
_____(2012b) 'Introduction', in J. Aston, B. Glynn and B. Johnson (eds) *Television, Sex and Society: Analyzing Contemporary Representations*, pp. ix–xxi.
Atkinson, H. (2012) *The Festival of Britain: A Land and Its People*, London: I.B.Tauris.
Attenborough, D. (1967) 'On Your Screen This Autumn', *Radio Times*, 30 September: 5.
_____(2002) *Life on Air: Memoirs of a Broadcaster*, London: BBC Books.
Baker Smith, M. (1950) 'The Festival of Britain 1951. The Television Pavilion', BBC WAC, R44/147, 13 April.
Bakhtin, M. (1968) *Rabelais and his World*, trans. H. Iswolsky, Cambridge, Mass.: Massachusetts Institute of Technology Press.
Bales, P. (1967) 'Proposed series: *European Holiday*', BBC WAC T14/2164/1, 9 November.
Barnes, G. (1951) 'Memo to all Department Heads', BBC WAC T14/436/5, 5 June.
Barthes, R. (1975) *The Pleasure of the Text*, trans. Richard Miller, New York: Hill and Wang.
_____(1981) *Camera Lucida: Reflections on Photography*, trans. Richard Howard, New York: Hill and Wang.
Batchelor, D. (2000) *Chromophobia*, London: Reaktion Books.
Bazin, A. (2003) 'Death Every Afternoon', trans. Mark A. Cohen, in I. Margulies (ed.) *Rites of Realism: Essays on Corporeal Cinema*, pp. 27–31.
BBC (1947) 'BBC Studio at Radiolympia', BBC WAC R44/500/5.
_____(1949a) 'Festival of Britain 1951 – BBC Participation [minutes]', BBC WAC R44/148, 21 November.
_____(1949b) 'Minutes of Radiolympia meeting, 2/5/49', BBC WAC R44/500/10.
_____(1949c) 'BBC Studio at Radiolympia', BBC WAC R44/500/11.
_____(1951a) 'Script for *Festival Close Up: Experiment in Design*, 21/5/51', BBC WAC T14/436/4.
_____(1951b) 'Beginners Please', BBC WAC R44/146/3.
_____(1952) 'Press Release', BBC WAC R44/498/5.
BBC General Advisory Council (1967) 'The Colour Service', June 16, BBC WAC file, R78/554/1.

Beech, D. (ed.) (2009a) *Beauty*, London and Cambridge, Mass.: Whitechapel and The MIT Press.

_____(2009b) 'Introduction: Art and the Politics of Beauty', in D. Beech (ed.) *Beauty*, pp. 12–19.

Beinart, W. and Coates, P. (1995) *Environment and History: The Taming of Nature in the US and South Africa*, London: Routledge.

Beinart, W. and McGregor, J. (eds) (2003) *Social History and African Environments*, Oxford: James Currey.

Bell, J. (1810) *Anatomy of the Bones, Muscles and Joints. With Plates. 1797*, London: Longman.

Belling, C. (1998) 'Reading *The Operation*: Television, Realism, and the Possession of Medical Knowledge', *Literature and Medicine*, 17, 1: 1–23.

Benjamin, W. (1999) 'The Work of Art in the Age of its Technological Reproducibility: Second version', in H. Elland and M. W. Jennings (eds) *Selected Writings Volume 3, 1935–1938*, pp. 101–33.

Bennett, James (2003) 'Public Service Broadcasting Goes Digital: Interactivity and Event Television (The Case of Natural History Programming)', paper presented at the Screen conference, University of Glasgow, 4–6 July.

_____(2008) 'Interfacing the Nation: Remediating Public Service Broadcasting in the Digital Television Age', *Convergence*, 14, 3: 277–99.

Bennett, Tony (1983) 'Really Useless Knowledge: A Political Critique of Aesthetics', *Thesis Eleven*, 12: 28–52.

Bevan, K. (2007) 'I Hate Myself for Watching Freak Show TV', *Guardian Online*, 3 April, http://www.theguardian.com/culture/tvandradioblog/2007/apr/03/ihatemyselffor-watchingfre, accessed 4 December 2014.

Bird Jr, W.L. (1999) *'Better Living': Advertising, Media, and the New Vocabulary of Business Leadership, 1935–1955*, Evanston, Ill.: Northwestern University Press.

Biressi, A. and Nunn, H. (2013) 'The London 2012 Olympic Games Opening Ceremony: History Answers Back', *Journal of Popular Television*, 1, 1: 113–120.

Bishop, H. (1949) 'Festival of Britain, 1951', BBC WAC R44/148, 15 July.

Boddy, W. (2004) *New Media and Popular Imagination: Launching Radio, Television, and Digital Media in the United States*, Oxford: Oxford University Press.

Bolter, J. D. and Grusin, R. (1998) *Remediation: Understanding New Media*, Cambridge, Mass.: MIT Press.

Bonner, F. (2005) 'Looking Inside: Showing Medical Operations on Ordinary television', in G. King (ed.) *The Spectacle of the Real: From Hollywood to Reality TV*, pp. 105–15.

_____(2013) 'Gaudy nights: Dance and Reality Television's Display of Talent', in J. Jacobs and S. Peacock (eds) *Television Aesthetics and Style*, pp. 251–67.

Boss, P. (1986) 'Vile Bodies and Bad Medicine', *Screen*, 27, 1: 14–25.

Bourdieu, P. ([1960] 1984) *Distinction*, trans. R. Nice, London and New York: Routledge.

Bousé, D. (2000) *Wildlife Films*, Philadelphia: University of Pennsylvania Press.

Brafield, W. (2011) 'Letter', *Radio Times*, 5 November: 156.

Braun, C. von (2000) '*Big Brother* oder Der frei zirkulierende Eros', *Tages-Anzeiger*, 5 December: 57.

Bibliography

Briggs, A. (1995) *The History of Broadcasting in the United Kingdom, Volume V: Competition*, Oxford: Oxford University Press.

Bronfen, E. (1992) *Over Her Dead Body: Death, Femininity and the Aesthetic*, London and New York: Routledge.

Brown, Maggie (2002) 'Media: Why Greg is the Wrong Man', *Guardian* (Media Section), 19 August: 2.

Brown, Mary Ellen (1987) 'The Politics of Soaps: Pleasure and Feminine Empowerment', *Australian Journal of Cultural Studies*, 4, 2: 1–25.

Brown, Simon and Abbott, S. (2010a) 'The Art of Sp(l)atter: Body Horror in *Dexter*', in Simon Brown and S. Abbott (eds) Dexter: *Investigating Cutting Edge Television*, pp. 205–20.

_____(eds) (2010b) *Dexter: Investigating Cutting Edge Television*, London: I.B.Tauris.

Browne, N. (1984) 'The Political Economy of the Television (Super) Text', *Quarterly Review of Film Studies*, 9, 3: 174–82.

Brunsdon, C. (1997) *Screen Tastes: Soap Opera to Satellite Dishes*, London and New York: Routledge.

_____(1998) 'Structures of anxiety: Recent British Television Crime Fiction', *Screen*, 39, 3: 223–43.

_____(2007) 'Towards a History of Empty Spaces', *Journal of British Cinema and Television*, 4, 2: 219–32.

_____(2013) 'Television Crime series, Women Police, and Fuddy-duddy feminism', *Feminist Media Studies*, 13, 3: 375–94.

Brunsdon, C., D'Acci, J. and Spigel, L. (eds) (1997) *Feminist Television Criticism: A Reader*, Oxford: Oxford University Press.

Brunsdon, C., Johnson, C., Moseley, R. and Wheatley, H. (2001) 'Factual Entertainment on British Television: The Midlands TV Research Group's "8–9 Project"', *European Journal of Cultural Studies*, 4, 1: 29–62.

Budge, B. (1998) 'Joan Collins and the Wilder Side of women: Exploring Pleasure and Representation', in L. Gamman and M. Marshment (eds) *The Female Gaze: Women as Viewers of Popular Culture*, pp. 102–11.

Bull, S. (2012) 'A Post-genomic Forensic Crime Drama: *CSI: Crime Scene Investigation* as cultural forum on science', Stockholm University PhD thesis.

_____(forthcoming, 2017) *New Genetics on Popular Television: The Rise of Test Tube TV*, Basingstoke: Palgrave MacMillan.

Buonanno, M. (2008) *The Age of Television: Experiences and Theories,* trans. J. Radice, Bristol: Intellect.

Burch, N. (1973) *The Theory of Film Practice*, London: Secker and Warburg.

Burgess, C. (2001) 'A Natural Relief', *Radio Times*, 29 September: 8.

Burke, E. ([1757] 1990) *A Philosophical Enquiry into the Origin of our Ideas of the Sublime and the Beautiful*, Oxford: Oxford University Press.

Caldwell, J. T. (1995) *Televisuality: Style, Crisis and Authority in American Television*, New Brunswick: Rutgers University Press.

Cameron, K. M. (1994) *Africa on Film: Beyond Black and White*, New York: Continuum.

Campbell, C. and Falk, P. (eds) (1997) *The Shopping Experience*, London: Sage.

Campney, G. (1961) 'The BBC at the Ideal Home Exhibition', BBC WAC R44/925/1, 6 February.

Cardiff, D. and Scannell, P. (1987) 'Broadcasting and National Unity', in J. Curran, A. Smith and P. Wingate (eds) *Impacts and Influences: Essays on Media Power in the Twentieth Century*, pp. 157–73.

Cardwell, S. (2013) 'Television aesthetics: Stylistic Analysis and Beyond', in J. Jacobs and S. Peacock (eds) *Television Aesthetics and Style*, pp. 23–44.

Carter, E. (2011) *Dietrich's Ghosts: The Sublime and the Beautiful in Third Reich Film*, Berkeley: University of California Press.

Cartwright, L. (1995) *Screening the Body: Tracing Medicine's Visual Culture*, Minneapolis and London: University of Minnesota Press.

Castle, T. (1993) *The Apparitional Lesbian: Female Homosexuality and Modern Culture*, New York: Columbia University Press.

Caughie, J. (2000) *Television Drama: Realism, Modernism and British Culture*, Oxford: Oxford University Press.

Caughie, J., Isaacs, J. and Matheson, M. (1986) 'Co-production in the Next Decade: Towards an International Public Service', in C. MacCabe and O. Stewart (eds) *The BBC and Public Service Broadcasting*, pp. 92–105.

Césaire, A. (2000) *Discourse on Colonialism*, trans. R. D. G. Kelley, London: *Monthly Review Press*.

Chamberlain, M.E. (1999) *Decolonization: The Fall of the European Empires* (2nd edn), Oxford: Blackwell.

Chapman, J. and Cull, N.J. (2009) *Projecting Empire: Imperialism and Popular Cinema*, London: I.B.Tauris.

Chivers, T. (2011) '*Frozen Planet*, BBC One, review', *Telegraph*, 26 October, http://www.telegraph.co.uk/culture/tvandradio/8851005/Frozen-Planet-BBC-One-review.html, accessed 3 October 2014.

Clark, D. L. and Myser, C. (1996) 'Being Humaned: Medical Documentaries and the Hyperrealization of Conjoined Twins', in R. Garland Thomson (ed.) *Freakery: Cultural Spectacles of the Extraordinary Body*, pp. 338–55.

Coates, W. (1951) 'Letter to Joanna Spicer, PA to the Director of Television', BBC WAC, T14/439, 10 February.

Coates, W. and Baker Smith, M. (1950) 'Festival of Britain 1951: Theme for Television Display in Pavilion South Bank 15', BBC WAC, T14/439, January–February.

Codell, J. (2011) 'Domesticating Empire in the 1930s: Metropole, Colony, Family', in L. Grieveson and C. MacCabe (eds) *Empire and Film*, pp. 189–203.

Collie, H. (2014) 'Television for women: Television, gender and the everyday', De Montfort University Thesis.

_____(2016) ' "I've Been Having Fantasies about Reagan and Carter Three Times a Week": Television's Role in Feminine Desire', in R. Moseley, H. Wheatley and H. Wood (eds) *Television for Women*.

Collins, N. (1950) 'Letter to Malcolm Baker Smith', BBC WAC, R44/147, 11 February.

Bibliography

Conekin, B. E. (2003) *'The Autobiography of a Nation': The 1951 Festival of Britain*, Manchester: Manchester University Press.

Conlan, T. (2005) 'More Body Talk for Channel 4', *Guardian* (Media Section), 29 August, http://www.theguardian.com/media/2005/aug/29/edinburghtvfestival2005.broadcasting, accessed 1 November 2014.

Cooper, D. E. (ed.) (1995) *A Companion to Aesthetics*, London: Blackwell.

Corner, J. (1999) *Critical Ideas in Television Studies*, Oxford: Oxford University Press.

Couldry, N. and McCarthy, A. (eds) (2004) *MediaSpace: Place, Scale and Culture in a Media Age*, London and New York: Routledge.

Coward, R. (2003) 'Perverts and Narcissists', *Guardian*, 1 January: 12.

Cox, I. (1951) *The South Bank Exhibition: A Guide to the Story It Tells*, London: HMSO.

Creeber, G. (2001) 'Cigarettes and Alcohol: Investigating Gender, Genre, and Gratification in *Prime Suspect*', *Television and New Media*, 2, 2: 149–66.

_____(ed.) (2015) *The Television Genre Book* London: BFI Palgrave.

Crisell, A. (2002) *An Introductory History of British Broadcasting* (2nd edn), London and New York: Routledge.

_____(2006) *A Study of Modern Television: Thinking Inside the Box*, Basingstoke: Palgrave Macmillan.

Crouch, D., Jackson, R. and Thompson, F. (eds) (2005a) *The Media and the Tourist Imagination: Converging Cultures*, London and New York: Routledge.

_____(2005b) 'Introduction: The Media and the Tourist Imagination', in D. Crouch, R. Jackson and F. Thompson (eds) *The Media and the Tourist Imagination: Converging Cultures*, pp. 1–13.

Crozier, I. (2010a) 'Introduction: Bodies in History – The Task of the Historian', in I. Crozier (ed.) *A Cultural History of the Human Body in the Modern Age*, pp. 1–22.

_____(ed.) (2010b) *A Cultural History of the Human Body in the Modern Age*, London: Bloomsbury.

Curran, Charles (1969) *Supporting a Public Service: A Speech Given by Charles J. Curran, Director General of the BBC, Radio Industries Club, 27 May 1969*, London: BBC Publishing.

Curran, James, Smith, A. and Wingate, P. (eds) (1987) *Impacts and Influences: Essays on Media Power in the Twentieth Century*, London and New York: Methuen.

Dalle Vache, A. and Price B. (eds) (2006) *Colour: The Film Reader*, London and New York: Routledge.

Davies, F. (2011) '*Embarrassing Bodies*: Public Sphere or Spectacle?', *Flow TV*, http://flowtv.org/2011/09/embarrassing-bodies/, accessed 13 December 2015.

Dayan, D. and Katz, E. (1992) *Media Events: The Live Broadcasting of History*, Cambridge, Mass.: Harvard University Press.

Deans, J. (2002) 'Channel 4's Autopsy Show Prompts Hundreds of Calls', *Guardian* (Media Section), 21 November, http://www.theguardian.com/media/2002/nov/21/broadcasting.channel4, accessed 1 November 2014.

_____(2004) 'BBC's Deep Blue Rides on a Wave of Success', *Guardian* (Media Section), 25 October, http://www.theguardian.com/media/2004/oct/25/broadcasting.film, accessed 31 August 2014.

Debnath, N. (2015) 'Aidan Turner Reveals the Secret Behind *Poldark's* Appeal', *Independent*, 29 March, http://www.independent.co.uk/news/people/news/aidan-turner-reveals-the-secret-behind-poldarks-appeal-10141910.html, accessed 18 May 2015.

Debord, G. (1995) *The Society of the Spectacle*, trans. Donald Nicholson-Smith, New York: Zone Books.

Denis, M. (1956) *A Leopard in My Lap*, London: Odhams Press.

_____(1960) *Ride a Rhino*, London: W.H. Allen & Co.

_____(1965) *At Home With Michaela Denis*, London: Hutchinson & Co.

Dijck, J. van (2002) 'Medical Documentary: Conjoined twins as a mediated spectacle', *Media, Culture and Society*, 24: 537–56.

_____(2005) *The Transparent Body: A Cultural Analysis of Medical Imaging*, Seattle: University of Washington Press.

Doane, M. A. (1982) 'Film and the Masquerade: Theorising the Female Spectator', *Screen*, 23, 3–4: 74–88.

Dodge, M. and Perkins, C. (2009) 'The "view from nowhere"? Spatial Politics and Cultural Significance of High-resolution Satellite Imagery', *Geoforum*, 40, 4: 497–501.

Dorling, J. (1946) 'Letter to Sir. William Haley [Director General of the BBC]', BBC WAC R44/500/3, 11 October.

Douglas, M. (1966) *Purity and Danger: An Analysis of the Concepts of Pollution and Taboo*, London: Ark Paperbacks.

Dovey, J. (2000) *Freakshow: First Person Media and Factual Television*, London: Pluto.

Dowell, B. (2008) '*Bodyshock* Beats *Clowns* in 9 pm Slot', *Guardian Online*, 8 April, http://www.theguardian.com/media/2008/apr/08/tvratings.television, accessed 1 November 2014.

Doye, M. (2005) 'Mediating Tourism: An Analysis of the Caribbean Holiday Experience in the UK National Press', in D. Crouch, R. Jackson and F. Thompson (eds) *The Media and the Tourist Imagination: Converging Cultures*, pp. 14–26.

Dunn, D. (2005) '"We Are Not Here to Make a Film about Italy, We Are Here to Make a Film about ME...": British Television Holiday Programmes' Representations of the Tourist Destination', in D. Crouch, R. Jackson and F. Thompson (eds) *The Media and the Tourist Imagination: Converging Cultures*, pp. 154–69.

Dyer, R. (1997) *White: Essays on Race and Culture*, London: Routledge.

_____et al. (eds) (1981) *Coronation Street*, London: BFI.

Easen, S. (2003) 'Film and the Festival of Britain 1951', in I. MacKillop and N. Sinyard (eds) *British Cinema in the 1950s: An Art in Peacetime*, pp. 51–63.

Eco, U. (2004) *On Beauty*, trans. A. McEwen, London: Secker and Warburg.

Elland, H. and Jennings, M.W. (eds) (1999) *Selected Writings Volume 3, 1935–1938*, Cambridge, Belknap.

Ellis, J. (1982) *Visible Fictions: Cinema, Television, Video*, London and New York: Routledge.

_____(2000) *Seeing Things: Television in the Age of Uncertainty*, London: I.B.Tauris.

Elsaesser, T. (1986) 'Film History as Social History: The Dieterle/Warner Brothers bio-pic', *Wide Angle*, 8, 2: 15–31.

Bibliography

Espiner, M. (2002) 'What to Say about *Body Worlds*', *Guardian*, 30 March: 15.

Everett, W. (ed.) (2007a) *Questions of Colour in Cinema: From Paintbrush to Pixel*, Bern: Peter Lang.

―――― (2007b) 'Mapping Colour: An Introduction to the Theories and Practices of Colour', in W. Everett (ed.) *Questions of Colour in Cinema: From Paintbrush to Pixel*, pp. 7–38.

Fickers, A. (2010) 'Techno-politics of Colour: Britain and the European struggle for a Colour Television standard', *Journal of British Cinema and Television*, 7, 1: 95–114.

Fieldhouse, D. (1996) 'For Richer, for Poorer?', in P. J. Marshall (ed.) *Cambridge Illustrated History of British Empire*, pp. 108–46.

Fisher, G. (1951) 'Archbishop of Canterbury's Closing Speech [Festival of Britain]', BBC WAC, R44/148, undated.

Fiske, J. (1987) *Television Culture*, London: Routledge.

Fiske, J. and Hartley, J. (1978) *Reading Television*, London and New York: Routledge.

Flitterman-Lewis, S. (1992) 'Psychoanalysis, Film, and Television', in R. C. Allen (ed.) *Channels of Discourse, Reassembled*, pp. 203–46.

Foreman, D. (1951) 'Future of Telecinema', *The Times*, 26 July: 7.

Foster, H. (1985) *Recodings: Art, Spectacle, Cultural Politics*, Seattle: Bay Press.

Foucault, M. ([1963] 2003) *The Birth of the Clinic*, trans. A. M. Sheridan Smith, London and New York: Routledge.

Fowler, C. and Helfield, G. (eds) (2006) *Representing the Rural: Space, Place, and Identity in Films About the Land*, Detroit: Wayne State University Press.

Foxton, M. (2002) 'Sliced Liver Anyone?', *Guardian*, 22 November: A5.

Frazer, A. (1957) 'Précis for the 1957 National Radio Show', BBC WAC R44/990/2, 8 July.

Friedlander, M. J. (1960) *On Art and Connoisseurship*, trans. T. Borenius, Boston: Beacon.

Frith, S. (2000) 'The Black Box: The Value of Television and the future of television research', *Screen*, 41, 1: 33–50.

Furness, H. (2015) 'BBC's #AskPoldark Descends into Farce as Aidan Turner Swamped with Indecent Proposals', *Telegraph*, 30 March, http://www.telegraph.co.uk/news/celebritynews/11503301/BBCs-Ask-Poldark-decends-into-farce-as-Aidan-Turner-swamped-with-indecent-proposals.html, accessed 18 May, 2015.

Furstenau, M. (ed.) (2010) *The Film Theory Reader: Debates and Arguments*, London and New York: Routledge.

Galt, R. (2011) *Pretty: Film and the Decorative Image*, New York: Columbia University Press.

Gamman, L. (1998) 'Watching the Detectives: The Enigma of the female gaze', in L. Gamman and M. Marshment (eds), *The Female Gaze: Women as Viewers of Popular Culture*, pp. 8–26.

Gamman, L. and Marshment, M. (eds) (1998) *The Female Gaze: Women as Viewers of Popular Culture*, London: The Women's Press.

Gamson, J. (1998) *Freaks Talkback: Tabloid Talk Shows and Sexual Nonconformity*, Chicago: University of Chicago Press.

Garland Thomson, R. (1996) *Freakery: Cultural Spectacles of the Extraordinary Body*, New York: New York University Press.

Gaskell, I. (2003) 'Beauty', in R. S. Nelson and R. Shiff (eds) *Critical Terms for Art History* (2nd edn), pp. 267–80.

Gault, B. and McIver Lopes, D. (eds) (2005) *The Routledge Companion to Aesthetics* (2nd edn), London and New York: Routledge.

Geraghty, C. (1981) 'The Continuous Serial – a Definition', in R. Dyer et al. (eds), *Coronation Street*, pp. 9–26.

———(2000) *British Cinema in the Fifties: Gender, Genre and the 'New Look'*, London: Routledge.

———(2003) 'Aesthetics and Quality in Popular Television Drama', *International Journal of Cultural Studies*, 6, 1: 25–45.

Gibbons, F. (2002a) 'Channel 4 Defiant on Staging "illegal" Autopsy', *Guardian*, 20 November: 6.

———(2002b) 'Anatomist Will Escape Charges', *Guardian*, 22 November: 7.

———(2003) 'Galleries Tap Rich Vein of Public's Lust for Blood', *Guardian*, 9 January: 5.

Gibson, O. (2012) 'Olympics Opening Ceremony: Danny Boyle Credits Late Father for Inspiration', *Guardian Online*, 27 July, http://www.theguardian.com/film/2012/jul/27/olympic-opening-ceremony-danny-boyle-father, accessed 13 October 2015.

Giddings, R. and Selby, K. (2001) *The Classic Serial on Television and Radio*, Basingstoke: Palgrave.

Giles, J. (2004) *The Parlour and the Suburb: Domestic Identities, Class, Femininity and Modernity*, Oxford and New York: Berg.

Gilroy, P. (2000) *Between Camps: Race, Identity and Nationalism at the End of the Colour Line*, London: Allen Lane.

Goodman, G. and Moseley, R. (2015) 'Why Academics Are Interested in the Male Body in *Poldark* and *Outlander*', *The Conversation*, 2 June, http://theconversation.com/why-academics-are-interested-in-the-male-body-in-poldark-and-outlander-42518, accessed 2 July 2015.

Grieveson, L. (2011) 'The Cinema and the (Common)Wealth of Nations', in L. Grieveson and C. MacCabe (eds) *Empire and Film*, pp. 73–113.

Grieveson, L. and MacCabe, C. (eds) (2011a) *Empire and Film*, London: Palgrave/BFI.

———(2011b) *Film and the End of Empire*, London: Palgrave/BFI.

Griffiths, A. (2002) *Wondrous Difference: Cinema, Anthropology and Turn-of-the-century Visual Culture*, New York: Columbia University Press.

Grindon, L. (1994) 'The Role of Spectacle and Excess in the critique of illusion', *Post Script*, 13, 2: 35–43.

Grosz, E. (1996) 'Intolerable Ambiguity: Freaks as/at the Limit', in R. Garland Thomson (ed.), *Freakery: Cultural Spectacles of the Extraordinary Body*, pp. 55–66.

Guinness, P. (1961) 'BBC Exhibit *Daily Mail* Ideal Home Exhibition 6 March – 3 April 1961', undated.

Gunning, T. (1986) 'The Cinema of Attractions: Early Film, its Spectator and the Avant-Garde', *Wide Angle*, 8, 3 & 4: 63–70.

Bibliography

Gunther, B. (2002) *Media Sex: What Are the Issues?*, Mahwah, N.J.: Lawrence Erlbaum Associates, Ltd.

Hale, M. (2012) 'On the Ice and Under it, Nature thrives', *New York Times*, 16 March, http://mobile.nytimes.com/2012/03/17/arts/television/frozen-planet-on-the-discovery-channel.html?_r=0, accessed 3 October 2014.

Hall, S. and Whannel, P. (1964) *The Popular Arts*, London: Hutchinson Educational.

Harris, P. and Connolly, K. (2002) 'Trade in Bodies is Linked to Corpse Art show', *Observer*, 17 March: 9.

Hartley, J. (1999) *Uses of Television*, London and New York: Routledge.

Hazell, F. (1951) 'Letter to Cecil McGivern, Controller of Programmes, Television)', BBC WAC T14/438, September.

Heller, D. (2011) '"Calling Out Around the World": The Global Appeal of Reality Dance formats', in T. Oren and S. Sharaf (eds) *Global Television Formats: Understanding Television Across Borders*, pp. 39–55.

Heritage, S. (2013) '*The Man with the Ten Stone Testicles*: A Triumph of Gross-out TV', *Guardian Online*, 25 June, http://www.theguardian.com/tv-and-radio/shortcuts/2013/jun/25/man-with-10-stone-testicles-tv, accessed 1 November 2014.

Hickethier, K. (1990) 'The Television Play in the Third Reich', *Historical Journal of Film, Radio and Television*, 10, 2: 163–86.

Hickey, D. (2009) 'Enter the Dragon: On the Vernacular of Beauty', in D. Beech (ed.) *Beauty*, pp. 22–30.

Higgins, K. M. (2009) 'Whatever Happened to Beauty? A Response to Danto', in D. Beech (ed.) *Beauty*, pp. 31–5.

Higson, A. (1984) 'Space, Place, Spectacle', *Screen*, 25, 45: 2–21.

⎯⎯⎯⎯ (2006) 'A Green and Pleasant Land: Rural Spaces and British Cinema', in Catherine Fowler and Gillian Helfield (eds) *Representing the Rural: Space, Place, and Identity in Films About the Land*, pp. 240–55.

Hill, J. and Church-Gibson, P. (eds) (1998) *The Oxford Guide to Film Studies*, Oxford: Oxford University Press.

Hilmes, M. (2008) 'Television Sound: Why the Silence?' *Music Sound and the Moving Image*, 2, 2: 152–61.

Hogarth, W. ([1753] 2009) *The Analysis of Beauty*, New York: Cosimo.

Hoggart, R. (1962) 'TV in a Free Society', *Contrast*, 2, 1: 6–11.

Holdsworth, A. (2006) '"Slow Television" and Stephen Poliakoff's *Shooting the Past*', *Journal of British Film and Television*, 3, 1: 128–33.

Holland, P. (2004) *Picturing Childhood: The Myth of the Child in Popular Imagery*, London: I.B.Tauris.

Holmes, S. (2014) '"You Don't Need Influence… All You Need is Your first opportunity!": The Early Broadcast Talent Show and the BBC', *Critical Studies in Television*, 9, 1: 23–42.

hooks, b. (1999) 'The Oppositional Gaze: Black Female Spectators', in S. Thornham (ed.) *Feminist Film Theory: A Reader*, pp. 307–20.

Hopkins, L. (2001) 'Mr Darcy's body: Privileging the Female Gaze', in L. Troost and S. Greenfield (eds) *Jane Austen in Hollywood*, pp. 111–21.

Horne, N. (2001) 'Editorial', *Radio Times*, 22 September: 7.

Inglis, F. (2000) *The Delicious History of the Holiday*, London and New York: Routledge.

Itzkoff, D. (2014) 'For *Game of Thrones*, Rising Unease Over rape's recurring role', *New York Times*, 2 May, http://www.nytimes.com/2014/05/03/arts/television/for-game-of-thrones-rising-unease-over-rapes-recurring-role.html?hp&_r=3, accessed 12 May 2015.

Jacobs, Jason (2000) *The Intimate Screen: Early British Television Drama*, Oxford: Oxford University Press.

_____(2001) 'Issues of Judgement and Value in Television Studies', *International Journal of Cultural Studies*, 4, 4: 427–47.

_____(2003) *Body Trauma TV: The New Hospital Dramas*, London: BFI.

_____(2009) 'Television's Illustrated Talks', *Critical Studies in Television*, http://cstonline.tv/televisions-illustrated-talks, accessed 12 December 2015.

Jacobs, Jason and Peacock, S. (eds) (2013) *Television Aesthetics and Style*, London: Bloomsbury.

Jacobs, Lea and de Cordova, R. (1982) 'Spectacle and Narrative Theory', *Quarterly Review of Film Studies*, 7, 4: 293–308.

Jaramillo, D.L. (2013) 'Rescuing Television From "the Cinematic": The Perils of Dismissing Television Style', in J. Jacobs and S. Peacock (eds) *Television Aesthetics and Style*, pp. 67–75.

Jeffries, S. (2002) 'The Naked and the Dead', *Guardian*, 19 March: A2.

Johnson, C. (2015) 'All Aboard Slow TV: Capturing the Experience of Walking on TV', *Critical Studies in Television Online*, 15 May, http://cstonline.tv/all-aboard-slow-tv-capturing-the-experience-of-walking-on-tv, accessed 23 July 2015.

Johnson, C. and Turnock, R. (eds) (2005) *ITV Cultures: Independent Television Over Fifty Years*, Maidenhead: Open University Press.

Jones, B. (2014) '*Game of Thrones*, sex and HBO: Where Did It Go Wrong For TV's Sexual Pioneers?', *Bent: A Queer Blog*, 2 June, http://blogs.indiewire.com/bent/game-of-thrones-sex-and-hbo-where-did-it-go-wrong-for-tvs-sexual-pioneers-20140602, accessed 12 May 2015.

Jowett, L. and Abbott, S. (2013) *TV Horror: Investigating the Dark Side of the Small Screen*, London: I.B.Tauris.

Juffer, J. (1998) *At Home with Pornography: Women, Sex and Everyday Life*, New York: New York University Press.

Kalmus, N. (1987) *The Critique of Judgement*, trans. W. Pluhar, Indianapolis and Cambridge: Hackett.

Kalmus, N. (2006) 'Colour Consciousness [1935]', in A. Dalle Vache and B. Price (eds) *Colour: The Film Reader*, pp. 24–9.

Kant, I. ([1790] 1960) *Observations on the Feeling of the Beautiful and the Sublime*, Berkeley: University of California Press.

Kaplan, E.A. (1997) *Looking for the Other: Feminism, Film and the Imperial Gaze*, New York: Routledge.

Bibliography

_____([1983] 2010) 'Is the Gaze Male?', in M. Furstenau (ed.) *The Film Theory Reader: Debates and Arguments*, pp. 209–21.

Katz, E., Gurevitch, M. and Haas, H. (1973) 'On the Use of the Mass Media for Important Things', *American Sociological Review*, 38: 164–81.

Kavka, M. (2008) *Reality Television, Affect and Intimacy: Reality Matters*, Basingstoke: Palgrave.

Keighron, P. (2002) 'What is the BBC For?', *Broadcast*, 13 December: 13.

Kellner, D. (2005) 'Media Culture and the Triumph of the Spectacle', in G. King (ed.) *The Spectacle of the Real: From Hollywood to Reality TV and Beyond*, pp. 23–36.

Kemp, M. and Wallace, M. (2000) *Spectacular Bodies: The Art and Science of the Human Body from Leonardo to Now*, London: Hayward Gallery Publishing.

King, Geoff (2000) *Spectacular Narratives: Hollywood in the Age of the Blockbuster*, London: I.B.Tauris.

_____(2005) *The Spectacle of the Real: From Hollywood to Reality TV*, Bristol: Intellect.

King, Norman (1984) *Abel Gance*, London, BFI Publishing.

Kingsbury, P. and Jones III, J.P. (2009) 'Walter Benjamin's Dionysian adventures on Google Earth', *Geoforum*, 40, 4: 502–13.

Kramer, P. (1998) 'Post-classical Hollywood', in J. Hill and P. Church-Gibson (eds) *The Oxford Guide to Film Studies*, pp. 289–309.

Kristeva, J. (1982) *Powers of Horror: An Essay on Abjection*, trans. L.S. Roudiez, New York: Columbia University Press.

Lawson, M. (2014) 'From Anatomy to Execution: The Problems of Portraying Death on TV', *Guardian Online*, 12 August, http://www.theguardian.com/tv-and-radio/tvandradioblog/2014/aug/12/the-beauty-of-anatomy-executed-death-on-tv, accessed 1 November 2014.

Lefebvre, M. (2006a) 'Between Setting and Landscape in the Cinema', in Martin Lefebvre (ed.) *Landscape and Film*, pp. 19–59.

_____(ed.) (2006b) *Landscape and Film*, London and New York: Routledge.

Lev, P. (ed.) (2003) *The History of American Cinema: Transforming the Screen, 1950–1959*, New York: Charles Scribner's Sons.

Leverette, M. (2008) ' "Cocksucker, Motherfucker, Tits" ', in M. Leverette, B. Lott and C. L. Buckley (eds) *It's Not TV: Watching HBO in the Post-Television Era*, pp. 123–51.

Leverette, M., Lott, B. and Buckley, C.L. (eds) (2008) *It's Not TV: Watching HBO in the Post-Television Era*, London and New York: Routledge.

Levy, A. (2005) *Female Chauvinist Pigs: Women and the Rise of Raunch Culture*, London: Pocket Books.

Lewis, R. and Foy, Y. (1971) *The British in Africa*, Worcester: The Trinity Press.

Linster, T. (2012) 'The London Olympics Opening Ceremony Includes a Lesbian Kiss Seen Around the World', *After Ellen*, 28 July, http://www.afterellen.com/tv/102520-the-london-olympics-opening-ceremony-includes-a-lesbian-kiss-seen-around-the-world, accessed 16 October, 2015.

Lloyd, John (1967) 'Memo: Treatment', BBC WAC T14/2164/1, 11 October.

Lloyd, Robert (2012) 'Television review: *Frozen Planet* on the Discovery Channel', *LA Times*, 16 March, http://articles.latimes.com/2012/mar/16/entertainment/la-et-0317-frozen-planet-20120315, accessed 3 October 2014.

Lury, K. (2005) *Interpreting Television*, London: Hodder Arnold.

Lyall, S. (2012) 'A Five-Ring Opening Circus, Weirdly and Unabashedly British', *New York Times*, 27 July, http://www.nytimes.com/2012/07/28/sports/olympics/in-olympic-opening-ceremony-britain-asserts-its-eccentric-identity.html?hp&_r=0, accessed 16 October 2015.

MacAloon, J. (ed.) (1984) *Rite, Festival, Spectacle, Game*, Chicago: University of Chicago Press.

McArthur, C. (1975/76) 'Days of Hope', *Screen*, 16, 4: 139–44.

McCabe, J. (2012) 'The Olympics: A British perspective', *Critical Studies in Television Online*, 6 September, http://cstonline.tv/olympics, accessed 16 October 2015.

MacCabe, C. and Stewart, O. (eds) (1986) *The BBC and Public Service Broadcasting*, Manchester: Manchester University Press.

McCarthy, A. (2001) *Ambient Television: Visual Culture and Public Space*, Durham, N.C.: Duke University Press.

McGivern, C. (1951) 'Letter to Dennis Foreman, Director of the BFI', BBC WAC T14/438, 19 September.

_____(1957) 'Memo: Scottish Radio Show', BBC WAC R44/1255/1, 7 June.

MacKenzie, J.M. (ed.) (1986a) *Imperialism and Popular Culture*, Manchester: Manchester University Press.

_____(1986b) *Propaganda and Empire: The Manipulation of British Public Opinion*, Manchester: Manchester University Press.

_____(1988) *The Empire of Nature: Hunting, Conservation and British Imperialism*, Manchester: Manchester University Press.

McKie, R. (2001) 'BBC Walks into a Storm Over Unnatural History Lessons', *The Observer*, 11 November: 11.

MacKillop, I. and Sinyard. N. (eds) (2003) *British Cinema in the 1950s: An Art in Peacetime*, Manchester: Manchester University Press.

McLean, F. (1967) *Colour Television (BBC Lunch-time Lectures Fifth Series – 6)*, London: BBC Publishing.

McMahon, J.A. (2005) 'Beauty', in B. Gault and D. McIver Lopes (eds) *The Routledge Companion to Aesthetics*, pp. 307–19.

McNair, B. (2002) *Striptease Culture: Sex, Media and the Democratisation of Desire*, London and New York: Routledge.

McNutt, M. (2011) 'Game of Thrones – "You Win or You Die"', *Cultural Learnings*, 29 May, http://cultural-learnings.com/2011/05/29/game-of-thrones-you-win-or-you-die/, accessed 12 May 2015.

Madden, C. (1954) 'Filming Wildlife in Africa', *Radio Times*, 22 October: 7.

Margulies, I. (ed.) (2003) *Rites of Realism: Essays on Corporeal Cinema*, Durham, N.C.: Duke University Press.

Marshall, P. J. (ed.) (1996) *Cambridge Illustrated History of British Empire*, Cambridge: Cambridge University Press.

Bibliography

Maugham-Brown, D. (1985) *Land, Freedom and Fiction: History and Ideology in Kenya*, London: Zed Books.

May, J. (1936) 'Christmas Tricks with the Microphone: How Does Television Progress? – and Another Gift from the BBC', *Ideal Home*, December: 517.

Maynard, R. (1974) *Africa on Film: Myth and Reality*, New Jersey: Hayden.

Medhurst, A. (2015) 'Of Limp-wristed Lions, Pink Hippos and Handbags: Facets of Kid Camp on British Children's Television', paper presented at the Story of Children's Television conference, University of Warwick, 6–7 July.

Mee, L., and Walker, J. (eds) (2014) *Cinema, Television and History: New Approaches*, Newcastle-upon-Tyne: Cambridge Scholars Press.

Meyrowitz, J. (1989) 'The Generalized Elsewhere', *Critical Studies in Mass Communication*, 6, 3: 326–34.

Mills, B. (2013) 'What Does it Mean to Call Television "Cinematic"?', in J. Jacobs and S. Peacock (eds) *Television Aesthetics and Style*, pp. 57–66.

Mills, S. (2005) *Gender and Colonial Space*, Manchester: Manchester University Press.

Mitchell, W. J. T. (ed.) (1994), *Landscape and Power*, Chicago: University of Chicago Press.

Moore, S. (1998) 'Here's Looking at You, Kid!', in L. Gamman and M. Marshment (eds) *The Female Gaze: Women as Viewers of Popular Culture*, pp. 44–59.

Morley, D. and Robins, K. (1995) *Spaces of Identity: Global Media, Electronic Landscapes and Cultural Boundaries*, London and New York: Routledge.

Moseley, R. (2000) 'Makeover Takeover on British television', *Screen*, 41, 3: 299–314.

_____(2013) ' "It's a Wild Country. Wild… Passionate… Strange": *Poldark* and the place-image of Cornwall', *Visual Culture in Britain*, 14, 2: 218–37.

Moseley, R., Wheatley, H. and Wood., H. (eds) (2016) *Television for Women*, London and New York: Routledge.

Mothersill, M. (1992) 'Beauty', in D. E. Cooper (ed.) *A Companion to Aesthetics*, 44–51.

Mulvey, L. (1975) 'Visual Pleasure and Narrative Cinema', *Screen*, 16, 3: 6–18.

Murray, S. (2005) *Hitch Your Antenna to the Stars: Early Television and Broadcast Stardom*, London and New York: Routledge.

Nava, M. (1997) 'Modernity's Disavowal: Women, the City and the Department Store', in C. Campbell and P. Falk (eds) *The Shopping Experience*, 56–91.

Neale, S. ([1983] 1992) 'Masculinity as Spectacle', in *Screen* (ed.) *The Sexual Subject: A Screen Reader in Sexuality*, pp. 277–87.

Nederveen Pieterse, J. (1992) *White on Black: Images of Africa and Blacks in Western Popular Culture*, New York: Yale University Press.

Negra, D. (2009) *What a Girl Wants? Fantasizing the Reclamation of Self in Postfeminism*, London and New York: Routledge.

Nelson, R. (2007) *State of Play: Contemporary 'High End' TV Drama*, Manchester: Manchester University Press.

Nelson, R. S. and Shiff, R. (eds) (2003) *Critical Terms for Art History* (2nd edn), Chicago and London: University of Chicago Press.

Newcomb, H. (ed.) (1994) *Television: The Cultural View*, Oxford: Oxford University Press.

Newcomb, H. and Hirsch, P. (1994) 'Television as Cultural Forum', in H. Newcomb (ed.) *Television: The Cultural View*, pp. 503–15.

Nicholls, C. S. (2005) *Red Strangers: The White Tribe of Kenya*, London: Timewell Press.

Nunn, H. and Biressi, A. (2003) '*Silent Witness*: Detection, Femininity and the Post-Mortem Body', *Feminist Media Studies*, 3, 2: 193–206.

O'Connor, D. (2010) 'Popular Beliefs', in I. Crozier (ed.) *A Cultural History of the Human Body in the Modern Age*, pp. 109–26.

Odone, C. (2002) 'Dead Ignorance', *Observer*, 24 November: 27.

Oren, T. and Sharaf, S. (eds) (2011) *Global Television Formats: Understanding Television Across Borders*, London and New York: Routledge.

Ormsby, A. (2012) 'London 2012 Opening Ceremony Draws 900 Million Viewers', *Reuters*, 7 August, http://uk.reuters.com/article/2012/08/07/uk-oly-ratings-day-idUKBRE8760V820120807?feedType=RSS&feedName=sportsNews, accessed 13 October 2015.

O'Rorke, I. (2001) 'Skinless Wonder...: Art or Anatomy', *Guardian*, 20 May: D5.

Ouellette, L. and Hay, J. (2008) 'Makeover Television, Governmentality and the Good Citizen', *Continuum*, 22, 4: 471–84.

Panos, L. (2015) 'The Arrival of Colour in BBC Drama and Rudolph Cartier's Colour Productions', *Critical Studies in Television*, 10, 3.

Parsons, T. H. (1999) '"Wakamba Warriors are Soldiers of the Queen": The evolution of the Kamba as a martial race, 1890–1970', *Ethnohistory* 46, 4: 671–701.

Patterson, M. (1961) 'BBC at the Ideal Home Exhibition, 1961 script', BBC WAC, T14/438, undated.

Perkins, C. (2014) 'Dancing on My Own: *Girls* and Television of the Body', *Critical Studies in Television*, 9, 3: 33–43.

Pidduck, J. (1998) 'Of Windows and Country Walks: Frames of Space and Movement in 1990s Austen adaptations', *Screen*, 39, 4: 381–400.

Plunkett, J. (2015) 'Gently Does it: From Canal Trips to Birdsong, BBC4 to Introduce "Slow TV"', *Guardian Online*, 1 May, http://www.theguardian.com/media/2015/may/01/gently-does-it-from-canal-trips-to-birdsong-bbc4-to-introduce-slow-tv, accessed 23 July 2015.

Postman, N. (1986) *Amusing Ourselves to Death*, London: Methuen.

Potter, L. (1951) 'Piccadilly Playback', BBC WAC R44/146/2, 30 May.

Pratt, M. L. (1992) *Imperial Eyes: Travel Writing and Transculturation*, New York: Routledge.

Price, B. (2006) '*Colour: The Film Reader* – General Introduction', in A. Dalle Vache and B. Price (eds) *Colour: The Film Reader*, pp. 1–9.

Proctor, J. (2001) 'Letter', *Radio Times*, 29 September: 6.

Pugh, M. (1954) 'Leopard-girl Says: I'll Talk Peace to Mau Mau', *Sunday Graphic*, 21 March: 7.

Radio Industry Council (R.I.C.) (1953) 'Minutes of Informal Liaison Meeting', BBC WAC R44/498/9, 16 November.

Radio Newsreel (1961) 'Ideal Home Exhibition, 1961', BBC WAC R44/925/1, 6 March.

Bibliography

Raeside, J. (2012) 'Is TV's Obsession with Embarrassing Ailments Unhealthy?', *Guardian Online*, 17 July, http://www.theguardian.com/tv-and-radio/tvandradioblog/2012/jul/17/tv-obsession-embarrassing-disease, accessed 2 November 2014.

Ralph, J. D. (1950a) 'Telecinema', BBC WAC, T14/438, 15 March.

———(1950b) 'Letter to Sir Henry French, British Film Producers Association', BBC WAC T14/438, 23 March.

Regis, A. K. (2012) 'Performance Anxiety and Costume Drama: Lesbian Sex on the BBC', in J. Aston, B. Glyn and B. Johnson (eds) *Television, Sex and Society: Analysing Contemporary Representations*, pp. 137–50.

Reeves, C. (ed.) (2010) *A Cultural History of the Human Body in the Enlightenment*, London: Bloomsbury.

Reynolds, S. (1967) 'Television', *Guardian*, 6 July, held in press clippings pack, BBC WAC file R78/555/1.

R.I.C. (1953) 'Cinema Size TV Screen', BBC WAC R44/498/9, 29 August.

Rice, T. (2011a) 'Exhibiting Africa: British Instructional Films and the Empire Series (1925–8)', in L. Grieveson and C. MacCabe (eds) *Empire and Film*, pp. 115–33.

———(2011b) 'From the Inside: The Colonial Film Unit and the Beginning of the End', in L. Grieveson and C. MacCabe (eds) *Film and the End of Empire*, pp. 135–53.

Richards, Jeffrey (1986) 'Boy's Own Empire: Feature Films and Imperialism in the 1930s', in J. M. Mackenzie (ed.) *Imperialism and Popular Culture*, pp. 140–64.

Richards, Neil (2015) 'The *Fifty Shades of Grey* Paradox', *Slate*, 13 February, http://www.slate.com/articles/technology/future_tense/2015/02/fifty_shades_of_grey_and_the_paradox_of_e_reader_privacy.html, accessed 15 May 2015.

Richardson, R. (2010) 'Popular Beliefs about the Dead Body', in C. Reeves (ed.) *A Cultural History of the Human Body in the Enlightenment*, pp. 93–112.

Ritchie, D. (1951) 'Interim Report on Piccadilly Exhibition', BBC WAC R44/146/3, 1 August.

Roberts, P. (2001) 'Letter', *Radio Times*, 6 October: 5.

Rodan, D. (2009) 'Large, Sleek, Slim, Stylish Flat Screens: Privatized space and the Televisual Experience', *Continuum: Journal of Media and Cultural Studies*, 23, 3: 367–82.

Rodowick, D. N. (1982) 'The Difficulty of Difference', *Wide Angle*, 5, 1: 4–15.

Rogers, K. (1951) 'Billing for Week 29 Festival Lights', BBC WAC, T14/436/6, 16 July.

Rony, F.T. (1996) *The Third Eye: Race, Cinema and Ethnographic Spectacle*, Durham, N.C.: Duke University Press.

Rowe, D. (1998) 'Why I let my husband die on television', *Sunday Mirror*, 26 April, http://www.thefreelibrary.com/WHY+I+LET+MY+HUSBAND+DIE+ON+TELEVISION%3B+ IT'S+ THE+MOST+CONTROVERSIAL...-a060653538, accessed 22 December 2014.

Rubin, M. (1993) *Showstoppers: Busby Berkeley and the Traditions of Spectacle*, New York: Columbia University Press.

Ryan, Deborah S. (1997) *Ideal Home Through the 20th Century:* Daily Mail *Ideal Home Exhibition*, London: Hazar Publishing.

Ryan, Maureen (2014a) '*Game Of Thrones* Controversial scene: Twelve Reasons it Matters', *Huffington Post*, 22 April, http://www.huffingtonpost.com/2014/04/22/game-of-thrones-rape_n_5192482.html, accessed 12 May 2015.

_____(2014b) '*Outlander*, the Wedding Episode and TV's Sexual Revolution', *Huffington Post*, 29 September, http://www.huffingtonpost.com/2014/09/29/outlander-wedding_n_5896284.html, accessed 18 May, 2015.

Sappol, M. (2010) 'Introduction: Empires in Bodies, Bodies in Empires', in M. Sappol and S.P. Rice (eds) *A Cultural History of the Human Body in the Age of Empire*, pp. 1–35.

Sappol, M. and Rice, S.P. (eds) (2010) *A Cultural History of the Human Body in the Age of Empire*, London: Bloomsbury.

Savage, T. (1968) 'Television Takes a Holiday', BBC WAC, T14/2164/1, undated.

Scanner, The (1938) 'Enter the Penumbrascope!', *Radio Times*, 24 June: 15.

Scarry, E. (1985) *The Body in Pain: The Making and Unmaking of the World*, Oxford: Oxford University Press.

_____(2009) 'On Beauty and Being Just', in D. Beech (ed.) *Beauty*, pp. 36–44.

Schama, S. (1995) *Landscape and Memory*, New York: Knopf.

Screen (ed.) (1992) *The Sexual Subject: A Screen Reader in Sexuality*, London and New York: Routledge.

Searle, A. (2002) 'Show of Dissected Bodies Cuts Both Ways', *Guardian*, 23 March: 11.

Sender, K. and Sullivan, M. (2008) 'Epidemics of Will, Failures of Self-Esteem: Responding to Fat Bodies in *The Biggest Loser* and *What Not to Wear*', *Continuum*, 22, 4: 573–84.

Shacklock, Z. (in prep.) 'Television Kinaesthetics', University of Warwick, unpublished thesis.

Shaw, C. (2006) 'The Lure of the Weird', *Guardian* (Media Section), 20 February, http://www.theguardian.com/media/2006/feb/20/broadcasting.comment, accessed 1 November 2014.

Siede, C. (2015) 'The Naked Hypocrisy of *Game Of Thrones*' Nudity', *BoingBoing*, 12 May, http://boingboing.net/2015/05/12/the-naked-hypocrisy-of-game-of.html, accessed 18 May 2015.

Skeggs, B. and Wood, H. (2012) *Reacting to Reality Television: Performance, Audience and Value*, London and New York: Routledge.

Sleeman, J.L. (1947) *From Rifle to Camera: The Reformation of a Big Game Hunter*, London: Jarrolds.

Smith, R. (2003) 'Double Jeopardy', *Guardian Online*, 9 December, http://www.theguardian.com/media/2003/dec/09/broadcasting.tvandradio, accessed 1 November 2014.

Smyth, R. (1979) 'The Development of British Colonial Film Policy, 1927–1939, with Special Reference to East and Central Africa', *Journal of African History*, 20: 437–50.

Spigel, L. (1988) 'Installing the Television Set: Popular Discourses on Television and Domestic Space, 1948–1955', *Camera Obscura*, 16: 9–46.

_____(1997) 'The Suburban Home Companion: Television and the Neighbourhood Ideal in Post-War America', in C. Brunsdon, J. D'Acci and L. Spigel (eds) *Feminist Television Criticism: A Reader*, pp. 211–34.

Bibliography

_____(2001) *Welcome to the Dreamhouse*, Durham, N.C.: Duke University Press.
Stanley, T. (2012) 'Danny Boyle's Olympic Opening Ceremony Was As ironic, Complex and Beautiful as Britain Herself', *Telegraph*, 28 July, http://blogs.telegraph.co.uk/news/timstanley/100173004/danny-boyles-olympic-opening-ceremony-was-as-ironic-complex-and-beautiful-as-britain-herself/, accessed 16 October 2015.
Steinberg, B. (2014) 'TV-News Outlets Grapple with How to Report on Grisly ISIS Videos', *Variety Online*, 3 September, http://variety.com/2014/tv/news/tv-news-outlets-grapple-with-how-to-report-on-grisly-isis-videos-1201297143/, accessed 19 December 2014.
Steinhart, E.I. (2006) *Black Poachers, White Hunters: A Social History of Hunting in Colonial Kenya*, Oxford: James Currey Ltd.
Stephens, E. (2012) 'Anatomical Imag(inari)es: The Cultural Impact of Medical Imaging Technologies', *Somatechnics*, 2, 2: 159–70.
_____(2013) *Anatomy as Spectacle: Public Exhibitions of the Body from 1700 to the Present*, Liverpool: University of Liverpool Press.
Stevens, C. (2015a), 'Exploring the Vid: A Critical Analysis of the Form and Its Works', University of Warwick unpublished PhD thesis.
_____(2015b) 'To watch *Wonder Woman*', *Feminist Media Studies*, 15, 5: 900–3.
Stewart, D. E. (1957) 'Scottish Radio and Television Exhibition, Kelvin Hall, Glasgow, May 22–June 1 1957', BBC WAC R44/1255/1, 13 June.
Stratton, J. and Ang, I. (1994) '*Sylvania Waters* and the Spectacular Exploding Family', *Screen*, 35, 1: 1–21.
Street, John (2000) 'Aesthetics, Policy and the Politics of Popular Culture', *European Journal of Cultural Studies*, 3, 1: 27–43.
Street, Sarah (2010) 'The Colour Dossier Introduction: The Mutability of Colour Space', *Screen*, 51, 4: 379–82.
_____(2012) 'Cinema, Colour and the Festival of Britain, 1951', *Visual Culture in Britain*, 13, 1: 83–99.
Streeton, W. L. (1951) 'Letter to J. D. Ralph', BBC WAC T14/438, 27 July.
Stulman Dennett, A. (1996) 'The Dime Museum Freak Show Reconfigured as Talk Show', in R. Garland Thomson (ed.) *Freakery: Cultural Spectacles of the Extraordinary Body*, pp. 315–26.
Sydney-Smith, S. (2007) 'Endless Interrogation: *Prime Suspect* Deconstructing Realism through the Female Body', *Feminist Media Studies*, 7, 2: 189–202.
Terrace, V. (1995) *Television Specials: 3,201 Entertainment Spectaculars, 1939–1993*, Jefferson, NC: McFarland.
Thomas, E. and Campney, G. (1960) 'Copy for the Ideal Home Exhibition Catalogue (1961)', BBC WAC R44/925/1, 15 December.
Thompson, F. (2010a) 'Is There a Geography Genre on British Television?: Explorations of the Hinterland from *Coast* to *Countryfile*', *Critical Studies in Television*, 5, 1: 57–68.
_____(2010b) '*Coast* and *Spooks*: On the Permeable National Boundaries of British Television', *Continuum: Journal of Media and Cultural Studies*, 24, 3: 429–38.
Thornham, S. (ed.) (1999) *Feminist Film Theory: A Reader*, Edinburgh: Edinburgh University Press.

_____(2003) 'A Good Body', *European Journal of Cultural Studies*, 6, 1: 75–94.

Thumim, J. (2004) *Inventing Television Culture: Men, Women and the Box*, Oxford: Oxford University Press.

Tomlinson, A. (1996) 'Olympic Spectacle: Opening Ceremonies and Some Paradoxes of Globalization', *Media, Culture and Society*, 18, 4: 583–602.

Treneman, A. (1998) 'Visual Arts: Vile Bodies?', *Independent Online*, 24 March, http://www.independent.co.uk/life-style/visual-arts-vile-bodies-1152137.html, accessed 19 December 2014.

Troost, L. and Greenfield, S. (eds) (2001) *Jane Austen in Hollywood*, Lexington: University of Kentucky Press.

Trout, J. (2014) '*Outlander* and the Female Gaze: Why Women are Watching', *Huffington Post*, 22 September, http://www.huffingtonpost.com/jenny-trout/outlander-and-the-female-_b_5859154.html, accessed 18 May 2015.

Tudor, A. (1995) 'Unruly Bodies, Unquiet Minds', *Body and Society*, 1, 1: 25–41.

Turner, B. (2011) *Beacon for Change: How the 1951 Festival of Britain Helped to Shape a New Age*, London: Aurum Press.

Turnock, R. (2007) *Television and Consumer Culture: Britain and the Transformation of Modernity*, London: I.B.Tauris.

University of Dundee (2009) 'Operating Theatre Design', Museum Services Website, http://www.dundee.ac.uk/museum/exhibitions/medical/operation/operation7/, accessed 11 December 2014.

Urry, J. (1990) *The Tourist Gaze: Leisure and Travel in Contemporary Societies*, London: Sage.

Usher, S. (1967) 'My Verdict on Colour TV', *Daily Sketch*, held in press clippings pack, BBC WAC file R78/555/1, 6 July.

Van Vuuren, L. (2003) 'The Many Myths of Laurens van der Post: Van der Post and Bushmen in the Television Series *Lost World of Kalahari* (1958)', *South African Historical Journal* 48, 1: 47–60.

Waldby, C. (1997) 'Revenants: The Visible Human Project and the Digital Uncanny', *Body and Society*, 3, 1: 1–16.

_____(2000) *The Visible Human Project: Informatic Bodies and Posthuman Medicine*, London and New York: Routledge.

Walker, P. (2009) 'Better Off at Home – Era of the Staycation Dawns as Britons Abandon Foreign Holidays', *Guardian*, 14 August: 3.

Wasko, J. (2003) 'Hollywood and Television in the 1950s: The Roots of Diversification', in P. Lev (ed.) *The History of American Cinema: Transforming the Screen, 1950–1959*, pp. 127–46.

Waters, M. (2012) 'Fangbanging: Sexing the Vampire in Alan Ball's *True Blood*', in J. Aston, B. Glynn and B. Johnson (eds) *Television, Sex and Society: Analyzing Contemporary Representations*, pp. 33–45.

Weber, B. (2005) 'Beauty, Desire and Anxiety: The Economy of Sameness in ABC's *Extreme Makeover*', *Genders*, 41.

Webster, W. (2011) 'Mumbo-jumbo, Magic and Modernity: Africa in British Cinema, 1946–65', in L. Grieveson and C. MacCabe (eds) *Film and the End of Empire*, pp. 237–50.

Bibliography

Weinstock, J. A. (1996) 'Freaks in space: "Extraterrestrialism" and "Deep-space Multiculturalism"', in R. Garland Thomson (ed.) *Freakery: Cultural Spectacles of the Extraordinary Body*, pp. 327–37.

West, R. (1965) *The White Tribes of Africa*, London: Jonathan Cape.

Whannel, G. (1984) 'Getting in the TV Frame', *The Listener*, 5 January: 29.

Wheatley, H. (2004) 'The Limits of Television? Natural History Programming and the Transformation of Public Service Broadcasting', *European Journal of Cultural Studies*, 7, 3: 325–39.

_____(2005) 'Rooms within Rooms: *Upstairs Downstairs* and the Studio Costume Drama of the 1970s', in C. Johnson and R. Turnock (eds) *ITV Cultures: Independent Television Over Fifty Years*, pp. 143–58.

_____(2006) *Gothic Television*, Manchester: Manchester University Press.

_____(ed.) (2007a) *Re-Viewing Television History: Critical Issues in Television Historiography*, London: I.B.Tauris.

_____(2007b) 'Introduction: Re-viewing Television Histories', in H. Wheatley (ed.) *Re-Viewing Television History: Critical Issues in Television Historiography*, pp. 1–12.

_____(2013) 'At Home on Safari: Colonial Spectacle, Domestic Space and 1950s Television', *Journal of British Cinema and Television*, 10, 2: 257–75.

_____(2014) '"Marvellous, Awesome, True-to-Life, Epoch-Making, a New Dimension": Reconsidering the Early History of Colour Television in Britain', in L. Mee and J. Walker (eds) *Cinema, Television and History: New Approaches*, pp. 142–63.

_____(2015a) '*Game of Thrones*', in G. Creeber (ed.) *The Television Genre Book* (3rd edn), pp. 60–1.

_____(2015b) 'Visual Pleasure and Narrative Television', *Feminist Media Studies*, 15, 5: 896–9.

_____(forthcoming, 2016) 'Television in the Ideal Home', in R. Moseley, H. Wheatley and H. Wood (eds) *Television for Women*.

White, M. (2004) 'The Attractions of Television: Reconsidering Liveness', in N. Couldry and A. McCarthy (eds) *MediaSpace: Place, Scale and Culture in a Media Age*, pp. 75–91.

Wiggin, M. (1967) 'Whitest of All on Colour TV', *Sunday Times*, 2 July held in press clippings pack, BBC WAC file R78/555/1.

Williams, C. S. (1968) 'BBC Colour Television', *Radio Times*, 11 January: 63.

Williams, Linda (1991) 'Film Bodies: Gender, Genre and Excess', *Film Quarterly*, 44, 4: 2–13.

_____(1999) *Hard Core: Power, Pleasure and the 'Frenzy of the Visible'* (2nd edn), Berkeley: University of California Press.

_____(2008) *Screening Sex*, Durham, N.C.: Duke University Press.

Williams, Linda Ruth (2005) *The Erotic Thriller in Contemporary Cinema*, Edinburgh: Edinburgh University Press.

Williams, Melanie (2015) '"Aren't We Naughty Mummies??!!": Adult Women's Fandom and Cbeebies' *Mr Bloom's Nursery*', paper presented at the Fan Studies Network conference, University of East Anglia, 27–8 June.

Williams, Raymond (1974) *Television: Technology and Cultural Form*, London: Fontana.

_____(1985) *The Country and the City*, London: Chatto and Windus.
Williams, Zoe (2012) 'TV Review', *Guardian Online*, 28 June, http://www.theguardian.com/tv-and-radio/2012/jun/28/baby-with-new-face-extraordinary, accessed 2 November 2014.
Winston, B. (1998) *Media Technology and Society: A History from the Telegraph to the Internet*, London and New York: Routledge.
Wiseman, E. (2010) 'Why Channel 4's *Embarrassing Bodies* is in Rude Health', *Observer Online*, 5 September, http://www.theguardian.com/tv-and-radio/2010/sep/05/embarrassing-bodies-channel-4-behind-the-scenes, accessed 2 November 2014.
Wolf, N. (2008) *Landscape Painting*, Cologne and London: Taschen.
Wollaston, S. (2007) 'Last Night's TV', *Guardian Online*, 8 April, http://www.theguardian.com/culture/tvandradioblog/2007/nov/14/lastnightstvtruestoriesth1, accessed 1 November 2014.
_____(2011) 'TV Review: *Frozen Planet*', *Guardian Online*, 26 October, http://www.theguardian.com/tv-and-radio/2011/oct/26/tv-review-frozen-planet, accessed 31 October 2014.
WWH (1938) 'Foreword' in *Ideal Home Exhibition 1938 Catalogue*, published by *Daily Mail*, p. 9.
Wyatt, J. (1994) *High Concept: Movies and Marketing in Hollywood*, Austin: University of Texas Press.

Index

N.B. Television genres are listed alphabetically, not as a separate sub-section.

Aaron, Michele 186–7, 188
Aaronovich, David 188
Abbott, Stacey 158
abjection 19, 153–4, 156–7, 165–6, 168, 174, 180–1, 185–6, 190
Abney and Teal (2011–12) 117
accidental erotic spectacle 191, 192, 205, 219–22
Adam, Kenneth 34, 67
Admags 139–40
advertising 11, 26–8, 47–8, 103, 106, 125–6, 139–40, 142, 204
African Patrol (1958–9) 86–7
Air Parade (1951) 233n
Airport 24/7: Miami (2012–) 142
Akutagawa, Ryūnosuke 69
Al Jazeera 178
Alexandra Palace 23–4, 31, 34, 40, 41
Ali, Barish 86
All Aboard! The Canal Trip (2015) 16–17
All-Coloured Variety, The (1947) 40–1
'Allo 'Allo (1982–92) 117
Amazon 9

Ambler, Charles 237n
America's Got Talent (2006–) 50
American Gladiators (1989–96) 6
anatomy 155, 161–9, 179, 181–9
Anatomy for Beginners (2005) 183–5
Andrejevic, Mark 157
Andrew, Dudley 10
Andrews, Dana 68
Andrews, Hannah 239n
Andrews, Malcolm 129
Andy Pandy (1950–2) 31
Ang, Ien 10
Ant and Dec's Saturday Night Takeaway (2003–) 13
Apprentice, The (2005–) 135
Armand and Michaela Denis (1957–8) 87, 236n
Armstrong, John 111, 115–16, 118, 238n
Arroyo, José 14
Art Screen TV 126
Arthurs, Jane 192–6, 207–8, 242n
Associated Rediffusion 44
Association of Cinematograph Television and Allied Technicians 36

Aston, James 195–6
Atkins, Ian 66, 69, 235n
Atkinson, Harriet 26, 45
Attenborough, David 59, 61, 64–8, 101–2, 105–7, 109, 120–1, 234n, 235n
Attenborough, Richard 102
Autopsy: The Last Hours Of… (2014) 183
Autopsy: Life and Death (2006) 183–5

Bachelor, The (2002–) 194
Badalamenti, Angelo 115–16
Bagpuss (1974) 117
Bahn TV In Fahrt (2003–8) 15
Baird Television (Development) Company 30, 37–8, 47, 233n
Baker Smith, Malcolm 30–2, 233n
Bakhtin, Mikhail 11
Bales, Peter 140–1
Banana (2015) 191, 202
Barthes, Roland 187, 190–1, 220
Batchelor, David 56–7, 72
Bazin, André 179

265

BBC 8, 9, 15–16, 23–4, 28–32, 34–6, 38–44, 48–9, 57, 59–75, 87, 99–101, 105–6, 108–9, 119–21, 123–4, 133, 134, 139–44, 213–14
  201 Piccadilly Exhibition 38–40
  White City site 32, 123–4
  see also individual programme titles
BBC2
  introduction of colour television 9, 57–75
  launch of 9, 139
Beaconsfield Productions 86
beauty 1, 10, 17–18, 20, 29, 68, 73, 86, 88, 97, 98–9, 106–7, 108–121, 123–4, 125, 127–8, 133, 135–6, 142, 146, 148, 168, 202, 226–7, 230–1, 238n
beauty fatigue 109, 121
Beauty of Maps, The (2010) 135–6
Becky Sharp (1935) 236n
Beijing Swings (2003) 179–80
Beinart, William 83, 237n
Belling, Catherine 241n
Below the Sahara (1953) 84–5
Benjamin, Walter 188, 207
Bennett, James 105
Bennett, Tony 100
Betjeman, John 127
Bevan, Kate 173
Beyer, John 188

Big Brother (2011–) 240n
Biggest Loser, The (2004–) 156–7
Billy Smart's Circus (1947–78) 13, 66
Bird Jr., William L. 38, 232n
Bird's Eye View (1969–71) 18, 122, 126–7
Biressi, Anita 157, 223
Black and White Minstrel Show (1958–78) 66
Blackadder Back and Forth (2000) 51
Bleak House (2005) 124
Blue Planet, The (2001) 99–109, 111, 114, 115, 119
Blue Planet Prom in the Park (2002) 51, 100–1, 105, 107–8, 234n
Boddy, William 25, 49–50, 129
body genres 19, 153–4
Body-Line (1937) 161–2
Bodyshock (2003–) 158–60, 170–1, 174–7, 241n
Body Story (1998) 163–4, 166
Body Worlds (exhibition) 182
Bolter, Jay David 166–7
Bonner, Frances 241n
Bonner, Paul 127
Boss, Pete 158
Bouquet of Barbed Wire (1976) 192
Bourdieu, Pierre 111, 118
Bousé, Derek 112
Boyle, Danny 223–4
Bradbury, Julia 131–2
Breaking Bad (2008–13) 115
Brideshead Revisited (1981) 101–2

Briggs, Asa 234n
Britain From Above (2008) 132, 136–7
Britain Presents America's Best (1957) 13
Britain's Favourite View (2007) 18, 122, 124, 129–30, 132, 134, 135
Britain's Got Talent (2007–) 13, 50
British Film Institute 34, 36–7, 233n
British Film Producers Association 36
British Instructional Films 84, 237n
British Radio Equipment Manufacturers' Association (BREMA) 63–4, 234n
Bronfen, Elisabeth 116, 180, 188–9
Brookside (1982–2003) 229
Brown, Simon 158
Browne, Nick 121
Brunsdon, Charlotte 52, 99, 101–2, 113–14, 131, 157
Budge, Belinda 212
Bull, Sofia 155, 157, 240n
Bullough, W. S. 162–3, 169
Buonanno, Milly 88–9, 155–6, 213
Burch, Noël 205
Burke, Edmund 113
Byers, Frank 116

Café Continental (1947–53) 40–2
Caldwell, John T. 2–7, 13, 17, 19, 65, 123, 154, 156, 164, 196, 201

Index

Cameron, Kenneth 84–5, 237n
Campney, George 48–9, 234n
Capon, Naomi 73
Cardiff, David 224
Cardwell, Sarah 238n
carnival 6, 11–13, 172
Carnival on Ice (1956) 12–13
Carter, Erica 110
Carter, John 141
Cartier, Rudolph 69, 73–4
Cartwright, Lisa 155, 160–1
Cassavetes, John 68
Castle, Terry 241
Castle Air 134
Casualty (1986–) 157–8
Caughie, John 105–6, 116, 238n
CBS 61; *see also individual programme titles*
Cellan Jones, James 70
Cerron, Milagros 160, 174–5
Césaire, Aimé 94
CGI 16, 19, 103, 126, 129, 134–7, 153, 164, 166–8, 187, 226
Channel 4 119, 176–7, 179, 182, 194; *see also individual programme titles*
Chapman, James 85–6, 237n
Charlie Drake Show, The (1967–68) 67
Chekov, Anton 69
Chicago Hope (1994–2000) 157–8
children's television 14, 31, 52, 117, 123, 166, 221–2, 242n

Chirurgenwerk (1990–) 165
Chivers, Tom 109, 121
chromophobia 56–8, 72
Clark, David L. 170
Clarke, George 52
Coast (2005–) 18, 122–4, 127, 130, 132–7, 142, 148, 239n
Coates, Peter 95, 237n
Coates, Wells 30, 32
Codell, Julie 237n
Collie, Hazel 193, 202, 209, 212, 221, 237n
Collins, Norman 30, 35
Colonial Film Unit 84
colonial television 80, 90, 95–6
colonialism 89–91
Colour Comes to Town exhibition 63–4
colour television 5, 13–14, 17, 57–75, 234–6n
competition 5, 9–10, 43–4, 170
Conekin, Becky E. 32
Conroy, David 70, 74, 235n, 236n
contact zone 79, 236n
contemplative viewing 1, 14, 18–19, 110, 123, 126–7, 130–3, 148, 190, 196, 206, 211, 215–16, 217, 230–1, 238n
Cookery with Philip Harben (1946–51) 40
cookery programmes 14, 30, 40, 52, 123, 135, 138, 149
Cool for Cats (1956–61) 234n

Corner, John 5, 10–11, 17, 19, 112
Coronation Street (1960–) 117, 222
Cottrell-Boyce, Frank 223
Countdown (1982–) 221
Court-Treatt, Major and Stella 84
Coward, Ros 179–80
Cradock, Fanny 52
Crane, Nicholas 132
Creeber, Glen 157
crime drama 114, 116, 157, 240n
Crisell, Andrew 63, 66, 170
Critical (2015–) 158
Crocker, Nicholas 96–7, 237n
Crossroads (1964–88) 222
Crouch, David 138, 145
Crozier, Ivan 154–5
Crucifixion (2012) 183
Cry Freedom (1987) 102
Cube, The (2009–) 13
Cucumber (2015) 191, 202, 222
Cull, Nicholas 85–6, 237n
Curran, Charles 65

Daily Mail 46, 141, 234n
Dangerous Journey (1944) 84
David (1951) 233n
Davies, Faye 241
Davies, Russell T. 202, 222
Dawn Chorus: The Sounds of Spring (2015) 16
Dayan, Daniel 224–6
Days of Hope (1975) 147
De Cordova, Richard 8
Deans, Jason 179

Debnath, Neelah 213–14
Debord, Guy 10
Denis, Armand and Michaela 18, 79, 83–97, 236–7n
Dijck, José van 153–5, 159–61, 163, 165–8, 170, 172, 182
Dimbleby, David 128, 132, 135
Dimbleby, Richard 36, 140
Discovery Channel 100, 105–6; *see also individual programme titles*
Doane, Mary Ann 205–6, 209–10
documentary television 6, 10, 15, 34, 62, 80, 84, 114, 117–18, 122, 138, 140, 170, 179, 186–7, 193, 202, 209, 233n, 236n, 240n, 241n; *see also* fascinoma documentaries, landscape television, medical documentaries, natural history programming, television autopsies, and individual programmes
Dodge, Martin 136
Done and Dusted 223
Dovey, Jon 169
Doye, Marcella 147
Dr Alice Roberts: Don't Die Young (2008) 240n
Dr Who (1963–) 166
Dr Who Prom (2008/2010/2013) 51

Dryden, Sir Noel 32
Dunn, David 143
DVD 60, 103, 123, 196
Dwarves in Showbiz (2002) 170
Dyer, Richard 91
Dying at Grace (2003) 186–7
Dyke, Greg 99, 101, 105

Easen, Sarah 30, 233n
Eastenders (1985–) 117
Ellis, John 2–3, 4, 19, 104, 123, 148, 155–6, 204, 206, 237n
Ellis, Robin 213
Elsaesser, Thomas 10
Embarrassing Bodies (2008–) 159, 169, 171, 176–7
Embarrassing Bodies: Kids (2010) 241n
Embarrassing Bodies: Live from the Clinic (2011–) 176–8
Embarrassing Bodies Down Under (2013) 241n
Embarrassing Fat Bodies (2010) 241n
Embarrassing Old Bodies (2010) 241n
Emergency Ward 10 (1957–67) 234n
Empire (1964) 232n
Empire Marketing Board 84
equity 36
ER (1994–2009) 157
erotics 9, 19, 119, 153, 190–3, 195–6, 198–9, 203–8, 210–12, 214–22, 229, 231, 232n, 241–2n
Espiner, Mark 182
Evanson, Steve 133, 239n
event television 16–17, 20, 47–8, 50–2, 66, 100–1, 182, 223–6, 230, 234n
Everett, Wendy 58–9
excess 8, 11, 65, 67, 71, 155–6, 164, 168, 175
Extraordinary People (2003–) 159, 170–3, 175–6, 240n
Eye on Research (1957–61) 240n

Family Portrait (1951) 233n
Fantastic Voyage (1966) 166
fanvids 217, 242n
fascinoma documentaries 6, 19, 153, 158–60, 169–78
Fast Track (2010–14) 142
female gaze 193, 207–19
Fenton, George 102, 105–6
Ferman, James 236n
Festival of Britain 17, 24, 25–40, 45, 46, 232–3n
Fickers, Andreas 234n
Fieldhouse, David 81
Fifty Shades of Grey (2011) 207
Filming Wild Animals (1954–5) 79, 87–8, 90–2, 94, 237n
Fiske, John 11–12, 155–6, 204, 206, 212

Index

Fixer, The (2008) 240
Fletcher, Melanie 223
Flitterman-Lewis,
    Sandy 204–5
Flog It (2002–) 134
flow 2–3, 99, 180, 191,
    220, 222
Flying Pictures 134
Forman, Denis 233n
Forward a Century
    (1951) 233n
Foster, Hal 10
Fothergill, Alastair
    109, 120–1
Foucault, Michel 158–60,
    162, 164, 167,
    169, 193
Fox, Sue 179
Foy, Yvonne 92
Frazer, Austin 43–4
freak show 159,
    169–73, 175–6
Freud, Sigmund 154, 199
Friedlander, Max J. 117–18
Frith, Simon 107
Frozen Planet (2011)
    109–10, 111,
    120–1, 238n
Furness, Hannah 214

Galt, Rosalind 67–8, 73
Game of Thrones (2011–)
    115–16, 148–9, 191,
    196–9, 201, 209, 218
Gamman, Lorraine
    212, 242n
Gandhi (1981) 102
Garland Thornton,
    Rosemarie 170
Gaskell, Ivan 111, 117
gaze, the 1–2, 26, 89–90,
    94, 126, 132, 137–8,

142–6, 153–6,
157–60, 167, 169,
172, 174–5, 176–8,
186, 189, 193, 197,
199–201, 203–6,
208–9, 211–19, 221–2
Geraghty, Christine
    85, 113, 118–19,
    212, 237n
Geraldo and his Orchestra
    (1946–7) 40
Get This! (1959–62) 140
Gibbons, Fiachra 182
Gibson, Owen 243n
Giddings, Robert 70
Gielgud, Val 32
gifs 217
Giles, David 70
Giles, Judy 27–8, 48
Gilroy, Paul 236n
Girls (2012–) 201–2, 242n
Gladiators (1992–2000)
    13, 203, 221
glance theory 2–5, 11,
    14–15, 155, 204–5,
    212, 222
Goodman, Gemma 218
Google Earth 126, 135–7
Goona Goona (1928) 84
Gorham, Maurice 34
Graham Norton Show, The
    (2007–) 13
Gray, Ann 20
Great British Bake Off, The
    (2010)
Great British Village Show,
    The (2007) 148–9
Grieveson, Lee 237n
Grindon, Leger 10
Grose-Krasne 86; *see*
    also individual
    programme titles

grotesque 12, 154, 158,
    168, 231
Grusin, Richard 166–7
Guinness, Perry 49
Gunning, Tom 5
Gunther, Barrie 202, 220
Gunther's E.R. (2007)
    183, 240n
Gurevitch, Michael 18

Haas, Hadassah 18
Hagens, Gunther von 159,
    161, 169, 179–80,
    181–6, 189, 241n
Hale, Mike 110
Haley, Sir William 40
Hall, Stuart 111
Hamilton, Hamish 223
Hampshire, Susan 70–2
Handmade (2015) 16
Hannibal
    (2013–15) 115–17
Happy Holidays
    (1956–7) 139
Harrington, Julia 175
Harris, Paul 241n
Harrison, Cassian 15–16
Hartley, John 17
Haskell, Molly 209
Hay, James 156–7
Hazell, Frank 33–5
HBO 119, 196; *see*
    also individual
    programme titles
Heritage, Stuart 240
heritage drama 14, 71,
    101, 105–6, 116, 123,
    125, 238n
Hesmondhalgh, Julie 202
Hickethier, Knut 233–4
High Chaparral
    (1967–71) 68

Higson, Andrew 239n
Hilmes, Michele 118–19
Hirsch, Paul 220
Hodgson, Patricia 99
Hogarth, William 113
Hoggart, Richard 98–9
Holdsworth, Amy 239n
Holiday (1969–2007)
    138, 140–4
Holiday: Heaven on Earth
    (2012) 142–3
Holiday Hijack (2011)
    143, 145–7
Holiday on Ice 1960
    (1960) 13
holiday programmes 6,
    18, 123,
    137–49
Holiday Showdown
    (2003–9) 143, 145–6
Holidays (1959) 139
Holidays Ahead
    (1957–8) 139
Holland, Patricia
    190–1, 198
Hollyoaks (1995–) 229
Holmes, Su 40
Honey We're Killing the Kids (2006–7) 157
hooks, bell 89
Hopkins, Lisa 213
Horizon (1964–) 163
Horne, Nigel 104
Horrible Histories Prom
    (2011) 51
How to Build a Human
    (2002) 240n
How Your Body Works
    (1958) 158, 162–3,
    169, 239n
Howard, Geoffrey 235n
Hulu 9

Human Body, The
    (1998) 158, 164–5,
    169, 187–8
Hussein, Saddam 181

I, Claudius (1976) 192
Ideal Home Exhibition
    (aka Ideal Home
    Show) 17, 24, 27–8,
    37–8, 40, 45–50,
    52–55, 232n, 234n
Impact (1966–8) 68
Imperial British East Africa
    Company 80–1
imperial gaze
    89–91, 93–4
Incredible Human Machine (2007) 158,
    164–6, 168, 240n
Independent Television
    Authority 44
Independent Television
    News 44
Independent Television
    Wonderland
    44–5, 48
Inglis, Fred 138
Inside the Human Body
    (2011) 158, 164,
    166–9, 187
Inside Nature's Giants
    (2009–) 183
intentional erotic
    spectacle 191–2, 214,
    218–9, 221–2
intentional landscape
    130–2, 143, 145, 147,
    149, 227
ISIS/ISIL 178–9
ITV 9, 24, 40, 43–4, 48, 57,
    60, 101, 119, 139–40,
    142, 236n, 237n;

see also individual
    programme titles
Itzkoff, Dave 242n

Jack Benny Spectacular
    (1959–60) 13
Jack Jackson Show, The
    (1955–9) 234
Jackson, Rhona 138, 145
Jacobs, Jason 41, 113,
    118–19, 157–8, 239n
Jacobs, Lea 8
James, E. L. 207
Jaramillo, Deborah L. 7
Jeffries, Stuart 183, 189
Jessen, Christian 169, 177–8
Jewel in the Crown
    (1984) 101
Jim's Inn (1957–63) 140
Joan Gilbert at Home
    (1952–4) 88
Joe Millionaire (2003)
Johnson, Catherine 52
Johnson, Martin
    and Osa 84
Jones, Bethany 197
Jones III, John Paul 136–7
jouissance 11
Jowell, Tessa 99, 105
Juffer, Jane 192,
    207–8, 218
Jungle Boy (1957) 86–7

Kalmus, Natalie 68
Kamba people 94, 237n
Kant, Immanuel
    110–11, 133
Kaplan, E. Ann 89, 94, 211
Katz, Elihu 18, 224–5
Kavka, Misha 156, 169
Kearton, Cherry 84
Keighron, Peter 101

Index

Kellner, Douglas 10
Kennedy, Ludovic 139
Kenya 18, 79–97, 112, 146–8, 236n, 237n
    in film 84–6
    history of 80–3, 92, 94–5, 236n
    National Parks in 82, 95
Kikuyu people 94, 237n
kinaesthesia 49, 153, 167, 185, 210, 239n, 242n
King, Allan 186
King, Geoff 1–2, 9
King, Norman 8
King Soloman's Mines (1950) 85
Kingsbury, Paul 136–7
Knick, The (2014) 158, 160
Knights and Warriors (1992) 6
Kramer, Peter 8
Kristeva, Julia 180, 185

Lady Chatterley's Lover (1993) 192
Lamb, Kenneth 66, 70, 235n
landscape 14, 16, 18, 68, 70, 73, 79–80, 82–4, 89–90, 94–6, 112, 115–16, 120–1, 122–49, 158, 164–8, 214, 227, 238–9n, 240n
    body as 158, 164–8, 240n
    in colonial history 80, 82–4, 147
    painting 122–3, 126, 128–9, 135
landscape television 6, 14, 17–19, 122–38, 142, 148–9, 227, 238–9n

Late Night Horror (1968) 73–4
Late Night Line Up (1964–72) 62
Lawson, Mark 181
Learoyd, Barry 60, 69, 235n
Lee, John 184
Lefebvre, Martin 16, 124, 130–2, 143
Leisure and Pleasure (1951–5) 139
Leopard in My Lap, A (1956) 93
Lev, Peter 9
Levy, Arial 194–5
Lewis, Roy 92
Lewis-Smith, Victor 176
Life Before Birth (1960) 163
light entertainment 12–14, 62, 66–7
Lights of London, The (1951) 29
Limits of Human Endurance, The (1952) 161
Linster, The 243n
Little Britain Live (2005–7) 51
Living Body, The (1972) 163
Llewellyn Bowen, Laurence 52–3
Lloyd, John 141
Lloyd, Robert 109, 111
Lost World of the Kalihari (1958) 236n
Lury, Karen 5, 63, 66, 155–6, 164
Lyall, Sarah 225
Lynam, Des 129

Maasai people 81, 94, 146–8
MacAloon, John 224
McArthur, Colin 147
MacCabe, Colin 147, 237n
McCabe, Janet 225, 229
McCarthy, Anna 24–5
McGivern, Cecil 35, 43
McGough, Roger 168
McGregor, JoAnn 83
MacKenzie, John M. 83, 236–7n
McLean, Francis 64–5
McMahon, Jennifer A. 110, 113
McNair, Brian 194–5
McNutt, Myles 197
Madden, Cecil 87
Mad Men (2007–15) 114, 116
Magic School Bus (1994–7) 166
Man Alive (1965–82) 240n
Marr, Andrew 132
Marshment, Margaret 212
Masters of Sex (2013–) 191, 199–201, 210–11, 221
Mau Mau 85, 94, 237n
Maugham-Brown, David 83
Maxim, Ernest 67
Medhurst, Andy 242n
medical documentaries 19, 153–5, 158–78, 181–9
medical gaze 158, 159–60, 167, 169, 176–8, 185–6
Men of Two Worlds (1946) 237
Mervyn, Richard 132–3, 134, 239n

271

Meyrowitz, Joshua 137–8
Michael, George and Marjorie 95, 237n
Michael Mosley: Infested! Living with Parasites 169
Michaela and Armand Denis (1955) 87, 236n
Michaels of Africa, The (1955–6) 237n
Milkshake Live (2008–) 51
Mills, Sara 79, 92
Miracles in the Womb (2007) 164, 168
Mirzoeff, Edward 127
Mitchell, W. J. T. 128
modernity 17, 24, 27–30, 45–8, 50, 53, 92
montage 16, 94, 106–7, 110, 116, 120, 123, 127–9, 133–4, 144, 146, 149, 164, 227–30, 239n, 243n
Moore, Harry 73–4
Moore, Suzanne 204
Morley, David 144
Morrison, David 107
Moscow State Circus (1959–61) 13
Moseley, Rachel 52, 218, 242n
Mosley, Michael 166, 169
Mothersill, Mary 111–13, 117
Motoring Holiday (1957–61) 139
Mowes, Herbie 187–8
Mr. Bloom 221–2
Mr. Tumble 221–3
Mulvey, Laura 12, 19, 154, 203–5, 208, 216, 242n

Murray, Susan 13
music hall 6, 13–14, 40, 42
My Shocking Story (2006–) 170–1, 173, 175
Myser, Catherine 170

Nairobi 81–2, 84, 95
Nasjonal vedkveldeight (2013) 15
national identity 128, 137, 148, 239n
National Radio Show 17, 24, 40–5, 48, 49, 51
National Viewers' and Listeners' Association 188
natural history programming 6, 18, 52, 79–111, 114–15, 117, 119–21, 123, 124, 237–8n
Natural History Unit (BBC) 96–7, 105–6
Nava, Mica 48
Nederveen Pieterse, Jan 86, 89
Negra, Diana 157
Nelson, Robin 191
Netflix 9
New Tricks (2003–) 240n
New York World's Fair 38, 232n
Newcomb, Horace 220
Nicholls, C. S. 81, 236n
Nick and Jessica Variety Hour, The (2004) 234n
Ninja Warrior UK (2015–) 13
No Passport (1960) 128, 140
North London Exhibition 232

Now TV 9
Nunn, Heather 157, 223

O'Connor, Dan 179
Olympic Broadcasting Services 223
Olympic Games, London (2012) 51, 223–4; *see also* Opening Ceremony of the London 2012 Olympic Games: Isles of Wonder (2012)
On Safari (1957–65) 79, 87–8, 92–3, 95–7
Once More with Felix (1967–70) 67, 235n
One Show, The (2006–) 240n
Only Fools and Horses (1981–2003) 117
Opening Ceremony of the London 2012 Olympic Games: Isles of Wonder (2012) 19–20, 223–31, 243n; *see also* Olympic Games, London (2012)
Ormsby, Avril 223
O'Rorke, Imogen 241n
Ouellette, Laurie 156–7
Outlander (2014–) 191, 208–9, 217–19
Outside Broadcast Unit 23, 28–9, 31, 34, 232–3n
Over the Hills (1957–63) 139
Owl Service, The (1969–70) 73, 236n

Index

Panos, Leah 74
Parsons, Timothy H. 237n
Passport (1960) 140
Patten, Brian 168
Perkins, Chris 136
Perkins, Claire 239n, 242n
Perry, Suzi 52
Perutz, Leo 39
Peters, Sylvia 39
Peyton Place (1957) 210–11
Peyton Place (1964–5) 210
Picture of Britain, A (2005) 18, 122, 127–8, 130, 132, 135, 137, 142, 148
Picture Page (1936–52) 31, 40
Pidduck, Julianne 238n
Pilkington Committee 98–9, 140
Planet of the Apes (1974) 221
Planet Earth (2006–8) 124
Poldark (1975–7) 213, 242n
Poldark (2015–) 191, 208–9, 213–16, 218
Poliakoff, Steven 132
pornography 19, 133–4, 153, 191–2, 193–6, 203, 207, 212, 242n
Portrait of a Lady (1968) 70–1, 235n
Portrait of a Marriage (1990) 192
Postman, Neill 10–11, 112
Potter, Leonard 39
Pratt, Mary Louise 79, 236n
Prehistoric Autopsy (2012) 183
Price, Brian 71

Pride and Prejudice (1995) 212–13
public operations 160
public service broadcasting 8–9, 18, 24, 34, 88, 98–106, 108–9, 119, 139, 163, 171, 177–8, 186, 202, 241n
Pullman, Jack 70

Queer as Folk (1999–2000) 202

Radio Rentals 63, 235n
Radio Times 66, 67, 70, 73, 103–4, 108–9, 233n
Radiolympia *see* National Radio Show
Radiolympia Follies (1949) 42
Raeside, Julia 176–7
Rainey, Paul 84
Rainey's African Hunt (1912) 84
Ralph, J. D. 34, 36–7
RDF Media 145; *see also* individual programme titles
reality television 117, 145–8, 156–7
Red Shoe Diaries, The (1992–7) 192
rediffusion 36–7
Regis, Amber K. 192
Reiner Moritz Productions 105
Remembering Summer (1959–60) 139
Replacements for Life (1960) 240n

Revenants, Les (2012–) 115
Reynolds, Stanley 60
Rice, Tom 237n
Richards, Jeffrey 80, 86, 237n
Richards, Neill 207
Richardson, Ruth 185–6
Ride a Rhino (1960) 93
Ritchie, Douglas 39
Robins, Kevin 144
Robinson, Anne 143
Rock and Rollergames (1989) 6
Rodan, Debbie 125–6
Rodowick, D. N. 211
Rogers, Keith 29
Rowe, David 188
Rubin, Martin 8
Ryan, Deborah S. 46, 234n
Ryan, Maureen 217, 242n

Safari (1956) 85
Santos, Rudy 172
Sappol, Michael 167, 175
Savage, Tom 142
Savage Splendour (1949) 84
Scannell, Paddy 224
Scarry, Elaine 112, 188
Schama, Simon 128
schools' television 49, 162–3, 169, 239n
science fiction 52, 107, 167, 240n
Science of Man, The (1963) 163, 240n
Science on Saturday: Human Biology (1961) 163
Scintillation! (1951) 12

scopophilia 154, 168, 170, 189, 190–1, 197, 203–5, 208, 217–18, 221, 242n
Scottish Radio and Television Exhibition 38, 43–4
Scruton, Roger 117, 238n
Searle, Adrian 241n
See It Now (1951) 15
'See Yourself on Television' exhibits 35, 37–40, 228
Selby, Keith 70
Sender, Katherine 156
Shacklock, Zoë 239
Shaftesbury, Lord Anthony Ashley Cooper 110
Shaw, Chris 173
'shiny floor' shows 6, 13–14, 50, 52, 224
Showcase (1960–2) 140
Showtime 196; *see also individual programme titles*
Siede, Caroline 197
Signoret, Simone 68
Simba (1955) 85, 237n
Simpsons, The (1989–) 166
Skeggs, Beverley 157
Skyfall (2012) 144–5
Skyworks 131, 134–5; *see also individual programme titles*
Sleep (1963) 232n
slow television 15–17, 126, 132–3, 239n
Smith, Andrew Hayden 202
Smyth, Rosaleen 237n
Songs of Praise (1961–) 134

Spigel, Lynn 47, 88, 129, 237n
sports programmes 66, 117, 203, 209, 221, 224, 229–30, 234n
Stampede (1930) 237
Stanley, Tim 225
Stars in Your Eyes (1946–50) 40–1
Starz 196; *see also individual programme titles*
Steinberg, Brian 178
Stephens, Elizabeth 155, 182–5
Stevens, Charlotte 217, 242n
Stewart, D. E. 38
Stratton, John 10
Street, John 100–1
Street, Sarah 26, 58
Streeton, W. L. 36
Strictly Come Dancing (2004–) 13
live tour 51
Struggle for Life (1935) 237n
Stulman Dennett, Andrea 170
Sullivan, Margaret 156
Sunday Night at the London Palladium (1955–74) 42, 234n
Superhuman (2000) 240n
Sutton, Shaun 71, 236n
Sydney-Smith, Susan 157, 240n

Take Me Out (2007–) 13
talent shows 13, 39–40, 45, 50
Tayler, Kathy 143–4

Telekinema 30–7, 233n
television
 405 line system 35
 625 line system 61, 63
 advertising of 26–8, 47–8, 125–6
 advertising on 11, 106, 139–40, 142, 204
 aerial filming 120, 122, 123, 127, 131, 133–5, 144, 226–8, 230, 239n
 of attractions 5, 14–15, 17
 autopsies 159–60, 181–6
 big screen 30–7, 41–4, 51, 54, 225–7, 233n, 243n
 cinematic 6–7, 104, 107
 circus 13, 66, 224
 colonial 18, 79–80, 86–97
 as colonising apparatus 18, 79–80, 88–9
 death on 9, 19, 153, 157–9, 178–89
 flat screen 19, 53, 125–6
 handheld screen 190, 206–7
 HD 2, 5, 14, 19, 61, 109, 115, 122–6, 128–30, 135
 horror 73–4, 108–9, 158, 166–8, 179, 181, 183–5, 188–9
 ice shows 12–13, 42
 music 224, 228–30
 news 15, 31, 36, 38, 44, 47, 62, 118, 144, 171, 178–9, 181, 194, 238n, 239n

Index

NTSC 59, 61–2
PAL 59, 61
quality 6, 7, 18–19, 98–108, 113–14, 119, 123, 127, 191, 195, 196
sex on 6, 11, 19, 73, 153, 191–203, 206, 211–12, 216–22, 241–2n
terrorists' use of 178–7
trash 6, 13
wall hanging 5, 19, 53, 125–6, 129
Television X 195–6
Tell Me You Love Me (2007) 191
Tennant, David D. 141
Terry Pratchett: Choosing to Die (2011) 187–8
Theater Television Network 234n
Theatre 625 (1964–8) 74
Theodore Roosevelt in Africa (1909) 84
Thirty Minute Theatre (1965–73) 74, 236n
This Week (1956–78) 234n
Thompson, Felix 122, 138, 145
Thompson, Kristen 71
Thornham Sue 185–6
Three Essays in Colour (1967) 69–70
Three of a Kind (1967) 60
Thumim, Janet 139
Thunderbirds (1965–6) 221
Time Out (1964–5) 139
Tipping the Velvet (2002) 192
Tobing Rony, Fatimah 93
Tofu (2015) 202

Tomlinson, Alan 10
Top of the Lake (2013–) 115
tourist gaze 137–8, 141–3, 145–8
Trailing Wild Animals (1923) 84
Travel Channel, The 142
Travel Show, The (2014–) 142, 144–5
Trout, Jenny 217
True Blood (2008–14) 191
True Detective (2014–) 115–16
Tucker, Rex 70
Turner, Aiden 213–14
Turner, Barry 32, 35
Turnock, Rob 239n
TV Frame 126
TV Times 44, 140
Twice Twenty (1955–8) 139
Twin Peaks (1990–1) 114–16

University of Warwick 226–7, 237–8n, 243n
Urry, John 137–8, 143, 145
Usher, Shaun 68
Utopia (2013–) 115

Val Parnell's Saturday Spectacular (aka *Saturday Spectacular*) (1956–61) 12
Vanity Fair (1967–8) 70–4, 234n, 236n
variety 6, 12–14, 23–4, 31, 37, 40–2, 44, 50, 66–7, 233–4n

vaudeville 6, 13, 23, 37, 41
Vicar's Wives (2007) 124
viewer research 100–1, 107–8, 114–19, 192–3, 195–6, 201–3, 206–10, 212, 216–17, 220–2
Vile Bodies (1998) 179
Virginian, The (1962–71) 68
Visible Human Project 161
visual pleasure 1–2, 4–5, 7–9, 11–2, 14, 17–20, 38, 50, 68, 70–3, 75, 95, 98–101, 103–12, 116, 118–19, 123–4, 126–7, 133–6, 143–4, 146–9, 153–6, 168, 170, 173–4, 190–2, 194–9, 201, 203–4, 209, 211–22, 224–7, 230–1, 238n
Voice, The (2012–) 13
voyeurism 157, 204–6, 212
Vuuren, Lauren Van 236n

Wagner (1983) 115
Wainwright, Alfred 131
Wainwright Walks (2007–9) 18, 122, 130–4
Waldby, Catherine 155, 161
Walking With Beasts (2001) 103
Walking With Beasts Live (2007–) 51
Wall, Sarah 243n
Wallace, Gregg 52, 54
Ward, Phil 67
Warhol, Andy 15, 232n

Warner Brothers 105
Warren Jr., Wesley 171, 175
Wasko, Janet 234n
Waters of Time (1951) 233n
Way We Travelled, The (2007) 140
Webster, Wendy 92–3, 236–7n
Weinstock, Jeffrey A. 170
West, Richard 81–2
Whannel, Gary 10, 111
What Not to Wear (2003–13) 156
Wheatley, Helen 20, 45, 52, 71, 73–4, 99, 197, 238n, 242n
Wheelbase (1964–74) 139
Whelpton, Peter 141
Where No Vultures Fly (1958) 236
Where Shall We Go? (1957–61) 139
Whicker, Alan 140

White, Mimi 2, 5, 14–15, 17
White Hunter (1957–8) 86
Who Wants to Marry a Millionaire (2000) 194
Why Beauty Matters (2009) 117, 238n
Wife Swap (2003–9) 145
Wiggin, Maurice 59, 63
Wild Cargo (1934) 84
wildlife programming *see* natural history programming
Williams, Linda 153, 157, 191–3, 203, 211–12
Williams, Melanie 221–2
Williams, Raymond 2, 13–14, 19, 116, 129, 163
Williams, Zoe 173–4
Wilson, Elizabeth 242n
Wilson, Emma 142–3
Wimbledon 48, 59–60, 64, 66, 235n
Winston, Brian 24

Winston, Robert 164–5, 169, 187–8
Winters, Shelley 68
Wiseman, Eva 176–7, 241n
Wish You Were Here? (1974–2003) 138–41
Wolf, Norbert 122–3, 129
Wollaston, Sam 110
Wonder of Man, The (1960) 163
Wood, Helen 157
Wood-Jones, Nik 131
World's Fair 38, 232n
Wyatt, Justin 8

X-Factor (2004–) 13
Tour 51

You Are What You Eat (2004–7) 157
Your Living Body (1969) 163
Your Own Time (1955–8) 139
YouTube 203, 205, 234n
Yu, Zhu 179–80, 241n